Yu-Jin Zhang
Image Engineering 1
De Gruyter Graduate

Also of Interest

Image Engineering Vol. 2: Image Analysis
Y-J. Zhang, 2017
ISBN 978-3-11-052033-0, e-ISBN 978-3-11-052428-4,
e-ISBN (EPUB) 978-3-11-052412-3

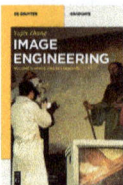

Image Engineering Vol. 3: Image Understanding
Y-J. Zhang, 2017
ISBN 978-3-11-052034-7, e-ISBN 978-3-11-052413-0,
e-ISBN (EPUB) 978-3-11-052423-9

Color Image Watermarking
Q. Su, 2016
ISBN 978-3-11-048757-2, e-ISBN 978-3-11-048773-2,
e-ISBN (EPUB) 978-3-11-048763-3, Set-ISBN 978-3-11-048776-3

Modern Communication Technology
N. Zivic, 2016
ISBN 978-3-11-041337-3, e-ISBN 978-3-11-041338-0,
e-ISBN (EPUB) 978-3-11-042390-7

Yu-Jin Zhang

Image Engineering

Volume I: Image Processing

DE GRUYTER

清华大学出版社
TSINGHUA UNIVERSITY PRESS

Author

Yu-Jin ZHANG
Department of Electronic Engineering
Tsinghua University, Beijing 100084
The People's Republic of China
E-mail: zhang-yj@tsinghua.edu.cn
Homepage: http://oa.ee.tsinghua.edu.cn/~zhangyujin/

ISBN 978-3-11-052032-3
e-ISBN (PDF) 978-3-11-052411-6
e-ISBN (EPUB) 978-3-11-052422-2

Library of Congress Cataloging-in-Publication Data
A CIP catalog record for this book has been applied for at the Library of Congress.

Bibliographic information published by the Deutsche Nationalbibliothek
The Deutsche Nationalbibliothek lists this publication in the Deutsche Nationalbibliografie; detailed bibliographic data are available on the Internet at http://dnb.dnb.de.

© 2017 Walter de Gruyter GmbH, Berlin/Boston
Typesetting: Integra Software Services Pvt. Ltd.
Printing and binding: CPI books GmbH, Leck
Cover image: Sorrentino, Pasquale/Science Photo Library
♾ Printed on acid-free paper
Printed in Germany

www.degruyter.com

Preface

This book is the Volume I of "Image Engineering," which is focused on "Image Processing," the low layer of image engineering.

This book has grown out of the author's research experience and teaching practices for full-time undergraduate and graduate students at various universities, as well as for students and engineers taking summer courses, in more than 20 years. It is prepared keeping in mind the students and instructors with the principal objective of introducing basic concepts, theories, methodologies, and techniques of image engineering in a vivid and pragmatic manner.

Image engineering is a broad subject encompassing other subjects such as computer science, electrical and electronic engineering, mathematics, physics, physiology, and psychology. Readers of this book should have some preliminary background in one of these areas. Knowledge of linear system theory, vector algebra, probability, and random process would be beneficial but may not be necessary.

This book consists of eight chapters covering the main branches of image processing. It has totally 55 sections, 99 subsections, with 164 figures, 25 tables, and 473 numbered equations, in addition to 60 examples and 96 problems (the solutions for 16 of them are provided in this book). Moreover, over 200 key references are given at the end of book for further study.

This book can be used for the first course "Image Processing" in the course series of image engineering, for undergraduate students of various disciplines such as computer science, electrical and electronic engineering, image pattern recognition, information processing, and intelligent information systems. It can also be of great help to scientists and engineers doing research and development in connection within related areas.

Special thanks go to De Gruyter and Tsinghua University Press, and their staff members. Their kind and professional assistance are truly appreciated.

Last but not least, I am deeply indebted to my wife and my daughter for their encouragement, patience, support, tolerance, and understanding during the writing of this book.

Yu-Jin ZHANG

Contents

1 Introduction to Image Processing

This book is the first volume of the book set "Image Engineering" and is focused on image processing.

Image processing has been developed for many years. In its broad sense, image processing comprises a number of techniques for treating images to obtain information from image. With the research and application of image technology, a more suitable term used is image engineering. Image engineering refers to a new discipline comprising all subjects relating to images, in which image processing (in its narrow sense) is an important part. This chapter provides an introduction to image processing, an overview of image engineering, and an outline of this book.

The sections of this chapter are arranged as follows:

Section 1.1 introduces the basic concepts and terms of image and its expression (including image representation and display form). A general representation of image is also discussed.

Section 1.2 provides a complete introduction to the field of image engineering. Image engineering currently includes three levels (image processing described in this book is the first level). It has a number of closely related disciplines, and has a wide range of applications. For a comprehensive understanding of the development and current situation of image engineering, some statistical data of image engineering literature, such as the number of papers published each year in a number of journals for the past more than 20 years, the categorization of there papers in 5 classes and further in 23 sub-classes, and the distribution of there papers in these classes and sub-classes, etc. are also presented and analyzed.

Section 1.3 describes the principal modules in the framework and system for image processing. This book mainly focuses on the principles, approaches, and algorithms of image processing. This section summarizes some peripheral knowledge and equipment of image processing, laying the foundation for later chapters to focus purely on image processing techniques.

Section 1.4 overviews the main contents of each chapter in the book and indicates the characteristics of the preparation and some prerequisite knowledge for this book.

1.1 Basic Concepts of Images

Images are an important medium by which human beings observe the majority of the information they receive from the real world.

1.1.1 Images and Digital Images

Images can be obtained by using different observing and capturing systems from the real world in various forms and manners. They can act, directly and/or indirectly, on

DOI 10.1515/9783110524116-001

human eyes and produce visual perception. The **human visual system** is a typical example of an observation system, from which humans perceive and recognize objects and scenes by observing and remembering their images.

The real world is three-dimensional (3-D) in space. However, the images obtained from the real word are generally two-dimensional (2-D). Such a **2-D image**, in most cases, is a 2-D light intensity function (2-D light pattern). An image can be represented by a 2-D array $f(x, y)$, where x and y are spatial coordinates and f is the amplitude at any pair of coordinates (x, y) representing certain properties of the scene projected on the image at that particular point. For example, in an image recording the brightness of scene, the amplitude f is proportional to the intensity of the image. In general, the amplitude f is proportional to one or several attributes of the image. For a real image, the values of x and y as well as f are real values, which are limited to a certain range.

Mathematically, an image can be explained as a function $f(x, y)$ with two variables. For the purpose of processing by computers, an **analog image** $f(x, y)$ should be digitalized to a **digital image** $I(r, c)$, in which r (row) and c (column) refer to the discrete position of any point in the digital image and the amplitude I refers to the discrete magnitude of the image at point (r, c). Since this book mainly discusses digital images, $f(x, y)$ is used to represent a digital image at any point (x, y) and f takes integer values only. A digital image can be considered a matrix whose rows and columns refer to the position of any point in the image and the corresponding matrix value refers to the intensity at that point.

In the early days, an image was called a "picture." When a picture is digitized, a sampling process is used to extract from the picture a discrete set of real numbers. The picture samples are usually quantized to a set of discrete gray-level values, which are often considered to be equally spaced. The result of sampling and quantizing is a digital picture. It is assumed that a digital picture is a rectangular array of integer values. An element of a digital picture is called a picture element (often abbreviated as **"pixel"** or **"pel"**). Although nowadays the term "image" rather than picture is used, because computers store numerical images of a picture or scene, the element of an image is still called a pixel. The element of a 3-D image is called a **voxel** (volume element). When combined, it is called an **imel** (2-D/3-D image element).

In its general sense, the word "image" refers to all entities that can be visualized, such as a still picture, a video, an animation, a graphic, a chart, a drawing, and text. These entities can be 2-D, 3-D, or even higher dimensions.

1.1.2 Matrix and Vector Representation

A 2-D image of $M \times N$ (where M and N are the total number of rows and the total number of columns of the image) can be represented either by a 2-D array $f(x, y)$ or by a 2-D matrix $\mathbf{F}$:

$$F = \begin{bmatrix} f_{11} & f_{12} & \cdots & f_{1N} \\ f_{21} & f_{22} & \cdots & f_{2N} \\ \vdots & \vdots & \ddots & \vdots \\ f_{M1} & f_{M2} & \cdots & f_{MN} \end{bmatrix} \tag{1.1}$$

The above matrix representation can also be converted to a vector representation. For example, the above equation can be rewritten as

$$F = \begin{bmatrix} f_1 f_2 \cdots f_N \end{bmatrix} \tag{1.2}$$

where

$$f_i = \begin{bmatrix} f_{1i} f_{2i} \cdots f_{Mi} \end{bmatrix}^{\mathrm{T}} \quad i = 1, 2, \cdots, N \tag{1.3}$$

Note that the array operation and matrix operations are different. Considering two 2×2 images $f(x, y)$ and $g(x, y)$ as an example, the array product is

$$f(x, y)g(x, y) = \begin{bmatrix} f_{11} & f_{12} \\ f_{21} & f_{22} \end{bmatrix} \begin{bmatrix} g_{11} & g_{12} \\ g_{21} & g_{22} \end{bmatrix} = \begin{bmatrix} f_{11}g_{11} & f_{12}g_{12} \\ f_{21}g_{21} & f_{22}g_{22} \end{bmatrix} \tag{1.4}$$

And their matrix product is

$$FG = \begin{bmatrix} f_{11} & f_{12} \\ f_{21} & f_{22} \end{bmatrix} \begin{bmatrix} g_{11} & g_{12} \\ g_{21} & g_{22} \end{bmatrix} = \begin{bmatrix} f_{11}g_{11} + f_{12}g_{21} & f_{12}g_{12} + f_{12}g_{22} \\ f_{21}g_{11} + f_{22}g_{21} & f_{21}g_{12} + f_{22}g_{22} \end{bmatrix} \tag{1.5}$$

1.1.3 Image Display

Image display means to put the image in a visible form that is shown on display equipment. In normal display equipment, an image is displayed as a 2-D pattern with different shades or colors.

Example 1.1 Digital image display
Figure 1.1 shows two examples of digital images. Two axis conventions are adopted in Figure 1.1 for image display. Figure 1.1(a) takes the top-left corner as the origin of the coordinate system, the row axis goes left-right, and the column axis goes top-down. Such a display is often used, for example, in screen displays. Figure 1.1(b) takes the

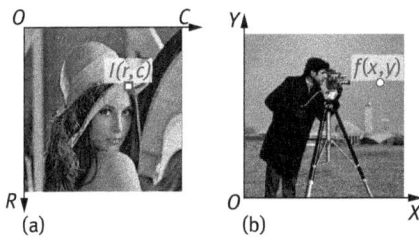

Figure 1.1: Examples of digital image display.

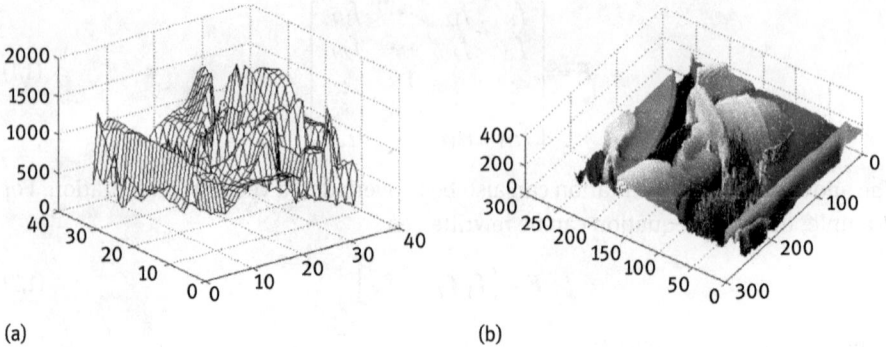

(a)   (b)

Figure 1.2: Other examples of digital image display.

down-left corner as the origin of the coordinate system, the X-axis goes left-right, and the Y-axis goes down-up. Such a display is used frequently, for example, in image manipulation.

In Figure 1.1, the gray levels of images represent the brightness of the image. Higher levels represent brighter pixels, while lower levels represent darker pixels. The gray levels of images can also be represented by a perspective plot in which the elevation corresponds to the brightness of pixel. One example of such a plot for Figure 1.1(a) is shown in Figure 1.2(a). A combination of the above two kinds of representations for Figure 1.1(a) is shown in Figure 1.2(b), in which the gray levels of images are represented by both elevation and brightness.

Example 1.2 Binary image representation and display
A binary image can be represented by a 2-D matrix, and there are different ways to show this matrix. Figure 1.3 gives three modes of showing a 4×4 binary image. In the mathematical model for image representation, the pixel area is represented by its center. The image thus obtained forms a discrete set of points on the plane, corresponding to Figure 1.3(a). If the pixel area is represented by its covered region, the Figure 1.3(b) is obtained. When the amplitude values corresponding to the pixel areas are marked in the image, the matrix representation results are shown as in Figure 1.3(c). The mode of Figure 1.3(b) may also be used for representing multiple grayscale images, where different shades of gray are required. The mode of Figure 1.3(b) may also be used for representing multiple grayscale images, in which case different gray values are represented by different number values.

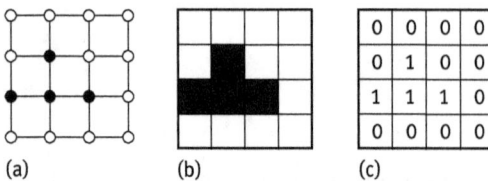

0	0	0	0
0	1	0	0
1	1	1	0
0	0	0	0

(a)   (b)   (c)

Figure 1.3: Three modes of representation for a 4×4 binary image.

1.1.4 High Dimensional Images

A basic representation of images is in the form of 2-D still gray-level images, which is represented by $f(x, y)$. A **general image representation function** is a vector function $f(x, y, z, t, \lambda)$ with five variables, where f stands for the properties of the world represented by the image, x, y, z are spatial variables, t is a time variable, and λ is a frequency variable (corresponding to wavelength). Some typical examples of the extension from $f(x, y)$ to $f(x, y, z, t, \lambda)$ are listed below.

(1) Consider that $f(x, y)$ is an image formed by the irradiance from object surfaces. If the object is cut along the capture direction into a number of sections and capture the image for each section, the integrated 3-D spatial information of the object (including its inside) can be obtained. In other words, a **3-D image** $f(x, y, z)$ is obtained. Imaging modalities, such as CT and MRI, are typical examples.

(2) Consider that $f(x, y)$ is a still image captured at a certain moment. If multiple images are captured consecutively along the time axis, the integrated 3-D temporal information (including dynamic information) of the object can be obtained. Video and other sequence images are examples of a 3-D image $f(x, y, t)$.

(3) Consider that $f(x, y)$ is an image which captured the irradiance of only one wavelength. If multiple wavelengths are captured, images with different properties (corresponding to reflection and absorption of different wavelengths λ) can be obtained. These images are either 3-D images $f(x, y, \lambda)$ or **4-D images** $f(x, y, t, \lambda)$. **Multi-spectrum images** are typical examples, in which each image corresponds to a certain wavelength, while all of them correspond to the same space and time.

(4) Consider that $f(x, y)$ is an image with only one property in the space location. In fact, the scene at one space location can have multiple properties. Therefore, an image can have several values at point (x, y) and it can be represented by a vector f. For example, a color image is an image having three RGB values at a pixel, $f(x, y) = [f_r(x, y), f_g(x, y), f_b(x, y)]$.

(5) Consider that $f(x, y)$ is an image obtained when projecting a 3-D scene onto 2-D plane. In this process, the depth or distance (from camera to object in scene) information in scene would be lost. If multiple images can be obtained from different viewpoints but for the same scene and if they can be combined, an image with all the information in the scene (including the depth information) can be obtained. The image with property as depth is called a **depth map:** $z = f(x, y)$. From the depth map, the 3-D image $f(x, y, z)$ can be derived.

1.2 Image Engineering

This book is focused on image processing that is part of image engineering. Several fundamental aspects of image engineering are given below.

1.2.1 Image Techniques and Image Engineering

Image techniques, which are expanding over wider and wider application areas, have attracted much attention in recent years. Because of the accumulation of solid research results and the improvement of electronic technology, many new theories have been proposed, many new techniques have been exploited, and many new applications have been created. It has become evident that a systematic study of the different branches of image techniques is of great importance.

In the following, a well-regulated explanation of the definition of image engineering, as well as its intention and extension, together with a classification of the theories of image engineering and the applications of image technology is provided.

Image engineering is a broad subject encompassing and based on subjects such as mathematics, physics, biology, physiology, psychology, electrical engineering, computer science, and automation. Its advances are also closely related to the development of telecommunications, biomedical engineering, remote sensing, document processing, industrial applications, and so on (Zhang, 1996c).

Some historical events related to image engineering made their effects a long time ago. Image techniques came into reality after the building of the first electronic computer ENIAC in 1946. The first image analysis started from grayscale data and ended with the derivation of a complete drawing in 1970. One of the earliest papers on understanding (line) drawings in a block world, such as making conclusions on what kinds of objects are presented, what relation exists between them, and what group they form, appeared in 1975. Note that this paper is included in a collected volume, *The Psychology of Computer Vision* (which introduces the first appearance of the word "Computer Vision (CV)" in literature).

The scope of image engineering has changed enormously since then. "Part of the reason for [this] change is that solid research results have been accumulated" (Kak, 1995). Many techniques have been developed, exploited, or applied only in the last decade. It can be seen that techniques for image engineering are implemented and used on a scale that few would have predicted a decade ago. It is also likely that these techniques will find more applications in the future (Zhang, 2008b; 2015c).

1.2.2 Three Levels of Image Engineering

What is the current "picture" of image engineering? Image engineering, from a perspective more oriented to techniques, should be referred to as a collection of the three related and partially overlapped image technique categories: image processing (IP), image analysis (IA), and image understanding (IU). In a structural sense, IP, IA, and IU build up three closely related and interconnected layers of IE as shown in Figure 1.4. Each layer operates on different elements and works with altered semantic levels. These layers follow a progression of increasing abstractness up and of decreasing compactness down.

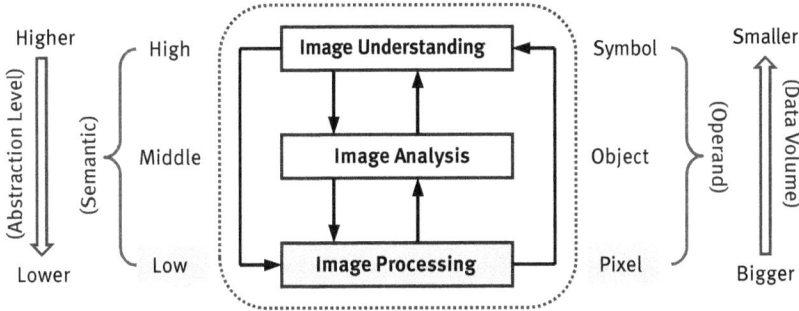

Figure 1.4: Three layers of image engineering.

Image processing primarily includes the acquisition, representation, compression, enhancement, restoration, and reconstruction of images. Although image processing concerns the manipulation of an image to produce another (improved) image, **image analysis** concerns the extraction of information from an image. In a general sense, it can be said that an image yields an image out for the former and an image yields data out for the latter. Here, the extracted data can be the measurement results associated with specific image properties or the representative symbols of certain object attributes. Based on image analysis, **image understanding** refers to a body of knowledge used in transforming these extracted data into certain commonly understood descriptions, and making subsequent decisions and actions according to this interpretation of the images.

One useful paradigm is to consider the three types of computerized operations in this continuum: low-level, mid-level, and high-level processes. Low-level processes involve primitive operations, such as image preprocessing to reduce noise, contrast enhancement, and image sharpening. A low-level process is characterized by the fact that both its inputs and outputs are images. Mid-level processing involves tasks such as segmentation, description of objects to reduce them to a suitable form for computer processing, and classification of individual objects. A mid-level process is characterized by the fact that its inputs are generally images, but its outputs are attributes extracted from those images. Finally, higher-level processing involves "making sense" of an ensemble of recognized objects, as in image analysis, and at the far end of the continuum, performing the cognitive functions normally associated with human visual system.

Example 1.3 Illustrative examples for IP, IA, and IU
One simple illustrative example, which pertains to image processing, image analysis, and image understanding, respectively, is given below.
(1) Image processing: Suppose an image is taken with insufficient light (as shown in Figure 1.5(a)), with suitable image processing techniques, a possible result would be Figure 1.5(b) with better clarity and contrast.

(2)   Image analysis: Continuing the above example, starting from Figure 1.6(a) (the result of image processing in Figure 1.5(b)), with suitable image analysis techniques, information can be extracted, as listed in Figure 1.6(b).

(3)   Image understanding: Continuing with the above example and starting from Figure 1.7(a) (the result of image analysis in Figure 1.6(b)), with suitable image understanding techniques, a few inferences and/or interpretations can be made, as listed in Figure 1.7(b).

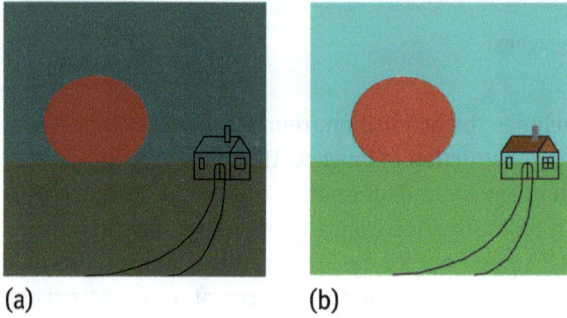

(a)   (b)

Figure 1.5: Input and output of image processing.

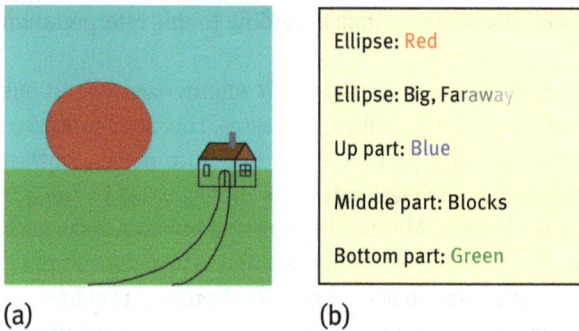

(a)   (b)

Ellipse: Red

Ellipse: Big, Faraway

Up part: Blue

Middle part: Blocks

Bottom part: Green

Figure 1.6: Input and output of image analysis.

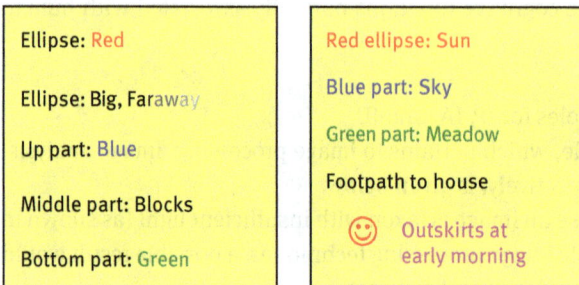

Ellipse: Red

Ellipse: Big, Faraway

Up part: Blue

Middle part: Blocks

Bottom part: Green

Red ellipse: Sun

Blue part: Sky

Green part: Meadow

Footpath to house

☺   Outskirts at early morning

Figure 1.7: Input and output of image understanding.

1.2.3 Related Disciplines and Fields

According to different science politics/perspectives, various terms such as **computer graphics** (CG), **pattern recognition** (PR), and **computer vision** (CV) are currently employed and (partially) overlapped with IP, IA, and/or IU. All of these subjects obtain help from a number of new techniques/tools, such as artificial intelligence (AI), fuzzy logic (FL), genetic algorithms (GA), neural networks (NN), soft science (SS), and wavelet transforms (WT). A diagram describing the relationship among the abovementioned subjects is given in Figure 1.8. Images are captured from the real world and processed to provide the basis for image analysis or pattern recognition. The former produces data that can be visualized by computer graphics techniques, while the latter continually classifies them into one of several categories that could be abstractly represented by symbols. Results produced by both can be converted to each other and can be further interpreted to help human beings understand the real world. This whole process aims to make computers capable of understanding environments from visual information, which is also the purpose of computer vision.

1.2.4 A Survey on Image Engineering

Image engineering is progressing very quickly. To capture the current status and the development trends of image engineering, a yearly survey series on image engineering has been on progress since 1996 and has lasted for more than 20 years (Zhang, 2015c; 2017).

In this survey series, the papers published in 15 important journals (covering electronic engineering, computer science, automation, and information technology) are considered. Those papers related to the image engineering are selected for classification. The papers are first classified into different categories, such as image processing, image analysis, and image understanding, then are further classified into different subcategories (see below), and finally, a number of statistics are gathered.

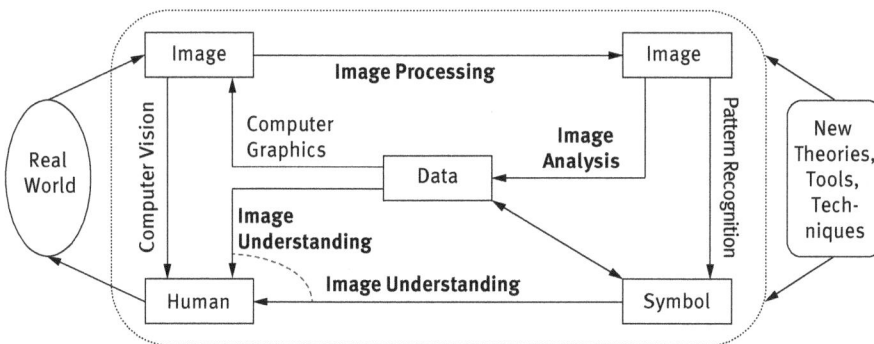

Figure 1.8: Image engineering and related subjects.

Table 1.1: Summary of image engineering over 22 years.

Year	#T	#S	SR	#IP
1995	997	147	14.7%	35 (23.8%)
1996	1,205	212	17.6%	52 (24.5%)
1997	1,438	280	19.5%	104 (37.1%)
1998	1,477	306	20.7%	108 (35.3%)
1999	2,048	388	19.0%	132 (34.0%)
2000	2,117	464	21.9%	165 (35.6%)
2001	2,297	481	20.9%	161 (33.5%)
2002	2,426	545	22.5%	178 (32.7%)
2003	2,341	577	24.7%	194 (33.6%)
2004	2,473	632	25.6%	235 (37.2%)
2005	2,734	656	24.0%	221 (33.7%)
2006	3,013	711	23.60	239 (33.6%)
2007	3,312	895	27.02	315 (35.2%)
2008	3,359	915	27.24	269 (29.4%)
2009	3,604	1,008	27.97	312 (31.0%)
2010	3,251	782	24.05	239 (30.6%)
2011	3,214	797	24.80	245 (30.7%)
2012	3,083	792	25.69	249 (31.4%)
2013	2,986	716	23.98	209 (29.2%)
2014	3,103	822	26.49	260 (31.6%)
2015	2,975	723	24.30	199 (27.5%)
2016	2,938	728	24.78	174 (23.9%)
Total	56,391	13,577		4,296 (31.64)
Average	2,563	617	24.08	195

A summary of the number of publications concerning image engineering from 1995 to 2016 is shown in Table 1.1. In Table 1.1, the total number of papers published in these journals (#T), the number of papers selected for survey as they are related to image engineering (#S), and the selection ratio (SR) for each year, as well as the numbers of papers belonging to image processing (#IP) are provided (the ratios of #IP/#S are also listed in parenthesis).

It is seen that image engineering is (more and more) becoming an important topic for electronic engineering, computer science, automation, and information technology. The SR reflects the relative importance of image engineering in the specialties covered by each publication. In 1995, the SR was only about 1/7. In 1996, the SR was about 1/6, then until 2002 it has been around 1/5 or so. The SR in 2003 was about one-fourth, and in 2004 the SR first exceeded one-fourth. In terms of the number of selected images, the number of selected image engineering documents in 2009 is nearly seven times that of 1995 (the total number of documents is only 3.6 times), which is remarkable considering the wide coverage of these journals. This results from the increase in the number of image engineering researches and contributors over the years; it is also the evidence of the vigorous development of image engineering disciplines.

Table 1.2: The classification of image processing category

Subcategory	# / Year
P1: Image capturing (including camera models and calibration) and storage	33
P2: Image reconstruction from projections or indirect sensing	18
P3: Filtering, transformation, enhancement, restoration, inpainting, quality assessing	58
P4: Image and/or video coding/decoding and international coding standards	45
P5: Image digital watermarking, forensic, image information hiding, etc.	42
P6: Multiple/super-resolutions (decomposition/interpolation, resolution conversion)	19

The selected papers on image engineering have been further classified into five categories: image processing (IP), image analysis (IA), image understanding (IU), technique applications (TA), and surveys.

In Table 1.1, the statistics for IP are also provided. The listed numbers are the numbers of papers published each year, and the percentages in parenthesis are the ratio of IP papers over all IE papers. Almost one-third papers related to image engineering are in the image processing area in these years. So far, image processing has achieved the most important progresses.

This survey series is characterized not only by categorizing the selected documents but also by statistically comparing and analyzing them, so that it is helpful for determining the direction of image engineering research and formulating the work decisions, in addition to contributing to the literature search.

The papers under the category IP have been further classified into six subcategories. The names of these six subcategories and the average numbers of papers per year for each subcategory are listed in Table 1.2.

1.3 Image Processing System

Image processing has been widely used. This book mainly introduces some basic techniques for image processing. Various image processing systems can be constructed by combining these techniques and can be used to solve different practical problems.

1.3.1 A Block Diagram of System

The constitution of a basic **image processing system** is represented in Figure 1.9. The seven modules shown in this figure have each a specific function, namely image acquisition (imaging), synthesis (image generation), processing, display, printing, communications, and storage. The system input includes image acquisition and generation, while the system output includes both display and printing. It should be noted that not every actual image processing system includes all of these modules. On the other hand, some special image processing system may also include other modules.

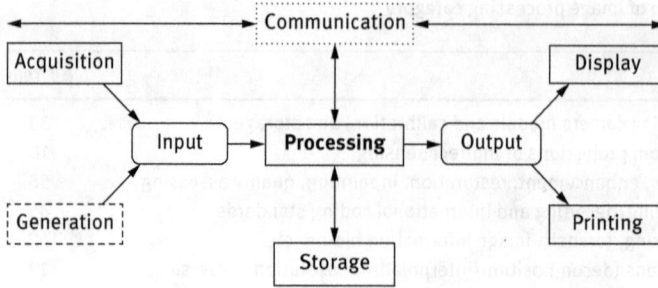

Figure 1.9: A configuration diagram showing an image processing system.

In Figure 1.9, the modules with dotted frames are associated with the image processing system, but the reader may refer to other books (see Further Reading at the end of this chapter) for introduction. The modules with solid frames, including the equipment needed for the respective functions, will be introduced briefly below.

1.3.2 Image Acquisition Equipment

Image acquisition is realized with certain equipment, also known as the imaging equipment, which transforms the objective scene into an image that could be processed by computer. In fact, the image can be collected from objective scene or generated (synthesized) from the known data. The latter is mainly the concern of computer graphics. For the image processing system, the common purpose of both is to get the discrete image that can be put into the computer for subsequent processing.

1.3.2.1 Acquisition Equipment Performance

The input of the image acquisition equipment at work is an objective scene, while the output is an image reflecting the nature of the scene. In general, there are two acquisition devices for image acquisition. One is a physical device (sensor) that is sensitive to the electromagnetic radiation energy of a spectral band (*e.g.*, X-ray, ultraviolet, visible, infrared), which generates the (analog) electrical signal proportional to the received electromagnetic energy, and the other is called a digitizer, which can convert the above (analog) electrical signal to a digital (discrete) form.

In the acquisition of visible light image, for example, the acquisition equipment projects the scene onto the image, which can be modeled by pinhole imaging in simple cases, so the objects in scene and their locations in image have corresponding geometry relationship. In addition, the brightness changes in scene should be reflected in the image, which requires the establishment of a certain illumination model, and links the scene brightness with the gray scale of images. The geometrical relation is finally associated with the **spatial resolution** of the image, whereas the illumination model is finally related to **amplitude resolution** of the image.

In addition to spatial resolution and amplitude resolution of the acquired image, the several indicators in the following are also often used to represent the properties of acquisition device (Young *et al.*, 1995).

(1) Linear response: It refers to whether the relationship between the intensity of input physical signal and the strength of output response signal is linear.

(2) Sensitivity: The absolute sensitivity can be represented by the minimum number of photons detected, while the relative sensitivity can be represented by the number of photons required to make the output to change one level.

(3) SNR: It refers to the (the energy or intensity) ratio of desired signal with unwanted interference (noise) in collected images.

(4) Unevenness: It refers to the phenomenon that physical input signal may be constant while the digital form of output is not constant.

(5) Pixel shape: It is the form of pixel region, which is generally square, but other shapes (*e.g.*, hexagonal) may also be possible.

(6) Spectral sensitivity: It is the relative sensitivity to radiation of different frequencies.

(7) Shutter speed: It corresponds to the acquisition shooting time.

(8) Read-out rate: It refers to the read (transmission) rate of signal data from the sensitive unit.

1.3.2.2 Solid-State Acquisition Devices

Solid-state array is a widely used collection device, which is composed of discrete silicon imaging elements, known as photosensitive primitive. Such a photosensitive element could produce the output voltage proportional to the input light intensity received. Solid-state array can be divided into two types according to the geometry organization: line scanner and flat-bed scanner. Line scanner comprises a row of photosensitive primitive. It relies on the relative motion between the scene and the detector to obtain 2-D images. The flat-bed scanner has photosensitive primitives lined square and can directly obtain 2-D images. A distinctive feature of solid planar sensor array is that it has a very fast shutter speed (up to 10^{-4} s), so many sports (with wild movement) can be frozen down. There are three commonly used devices.

CCD Device   Camera with **charge-coupled device** (CCD) is one of the most widely used image acquisition device. Its main component, that is, CCD sensor, is a solid-state imaging device that uses the mode of charge storage and transfer as well as readout for work, whose output is converted to a digital image by the digitizer (achieved by inserting a special hardware card in the computer).

The number of units in a CCD sensor array is 768×494 in the US standard (US RS-170 video norm), 756×582 in the European standard (European CCIR norm). According to different chip size, the unit size is often between in $6.5\,\mu m \times 6\,\mu m$ to $11\,\mu m \times 13\,\mu m$.

CCD sensor's advantages include the following: It has precise and stable geometry. It is small in size, high in strength, and resistant to vibration. It has high sensitivity (especially cooled to a lower temperature). It can be made of many resolutions and frame rates. It is able to imaging with invisible radiation.

CMOS Device Complementary metal oxide semiconductor (CMOS) sensor includes a sensor core, an analog-digital convertor, output registers, control registers, and gain amplifiers. The photosensitive pixel circuits in sensor core can be classified into three kinds. One is photodiode-type passive pixel structure, which is formed by a reverse biased photodiode and a switch tube. When the switch tube is turned on, the photodiode is connected to the vertical column lines, the amplifier located at the end of the column line reads out column line voltage. When the photodiode stored signal is read, the voltage is reset, then the amplifier transforms the charge proportional to optical signal into voltage output. Another is photodiode-type active pixel structure, which has more than the passive pixel structure an active amplifier on the pixel unit. The last one is raster active pixel architecture. The signal charges are integrated in raster, the diffusion point was reset before output. Then, the raster pulses are changed, and the raster signal charges are collected and transferred to the diffusion points. The voltage difference between the reset level and the signal voltage level is the output signal.

Compared with the conventional CCD imaging devices, CMOS imaging device integrates the entire system on a single chip, which reduces the power consumption and device size, and the overall cost is lower.

CID Device Charge-injection device (CID) sensor has an electrode matrix corresponding to image matrix. There are two isolated electrodes to produce potential wells at each pixel location. One electrode is connected to all the electrodes (and their corresponding pixels) in the same row, and the other electrode is connected to all the electrodes (and their corresponding pixels) in the same column. In other words, in order to access a pixel, its rows and columns should be selected.

The voltages of the above two electrodes can be respectively positive and negative (including zero), and there are three kinds of situations for their combination with each other, corresponding to three working modes of CID. The first is integration mode. In this case the voltage of the two electrodes are both positive, optoelectronics will accumulate. If all the rows and columns remain positive, the entire chip will give an image. The second is non-elimination mode. In this case the voltages of the two electrodes are negative and positive, respectively. The photoelectrons accumulated by the negative electrode can be migrated to the positive electrode and will inspire a pulse in the circuits connected to the other electrode, whose magnitude reflects the number of photoelectrons accumulated. The photoelectrons migrated will remain in the potential well, so the pixels can be read out repeatedly without removing them through a round-trip migration of charges. The third one is an elimination mode. In

this case the voltages of the two electrodes are both negative. The accumulated photoelectrons will overspill or inject into a silicon chip layer between the electrodes, and
this will excite pulses in the circuit. Similarly, the amplitude of the pulse reflects the
cumulative number of photoelectrons. However, this process will exclude migrated
photoelectrons out of potential well, so it can be used to "clearing" to make a chip
ready to capture another image.

The circuit in the chip controls the voltages of row and column electrodes to
acquire an image and to read out with the elimination mode or the non-elimination
mode. This allows the CID to access each pixel in any order, and at any speed in
reading sub-image of any size.

Compared with the conventional CCD imaging device, CID imaging device has
much lower light sensitivity. On the other side, it has the advantages of random access,
no blooming, etc.

1.3.3 Image Display and Printing

For image processing, the result of processing is still an image and is mainly used to
display for human viewing. So the **image display** is very important for the image processing system. The image display is an important step in the communication between
the system and the users.

The commonly used display devices in image processing systems include the
cathode ray tube (CRT), liquid crystal display (LCD) monitors and TVs. In the CRT,
the horizontal and vertical positions of the electron beam gun are controlled by the
computer. In each deflected position, the intensity of the electron beam gun is mod-
ulated by the voltage. The voltage at each point is proportional to the corresponding
grayscale value at this point. In this way, the grayscale image is converted to a brightness change mode, and this mode is recorded on a cathode ray tube screen. The
images entering the monitor can also be converted to hard copy slides, photos, or
transparencies.

In addition to the monitor, various printing devices, such as a variety of printers,
can also be considered as an image display device. Printers are generally used for low-
resolution output image. A simple method of early printing grayscale images on paper
is to use the repeated printing capabilities of standard-line printers; the gray value at
any point in the output image is controlled by the number of characters and the density of the dot printing. Printers that are currently in use, such as thermos-sensitive,
dye-sublimation, ink jet, and laser printers, can print higher-resolution images.

1.3.3.1 Halftone Output

Some printers can output now continuous tone at the same spatial position, such
as continuous-tone dye-sublimation printers. However, because of their slower speed
and the requirement of special materials (special paper and ink), their applications

have certain limitations. Most print devices can only directly output binary image, such as a laser printer whose output has only two gray scales (black for print output and white for no-print output). To output the grayscale image on a binary device and to keep their original gray level, a technique called halftone output should be used.

Halftone output technology can be viewed as a transforming technology from grayscale image into a binary image. It intends to convert all kinds of gray values into binary dot patterns, which can be output by the printing devices that can only directly output binary point. In the meantime, it also utilizes the integrated nature of the human eye, by controlling the output in the form of binary dot pattern (including the number, size, and shape) to obtain a visual sense of multiple gray levels. In other words, the output image produced by halftone technology is still the **binary image** in a very fine scale, but is perceived as grayscale image in the coarse scale, because of local average effect of the eye. For example, in a binary image each pixel only takes white or black, but from a distance, the human eye can perceive a unit composed of multiple pixels, then the human eye perceives the average gray value of this unit (which is proportional to the number of black pixels).

Halftone technology is mainly divided into two categories: amplitude modulation and frequency modulation. They are introduced in the following sections.

1.3.3.2 Amplitude Modulation

Initially proposed and used halftone output technology is called **amplitude modulation** (AM) halftoning, which adjusts the output size of the black dot to show the different shades of gray. The gray level of early pictures on newspapers is represented by using different sizes of ink dots. When observed from a distance, a group of small set of ink dots could produce some light gray visual effects, and a group of large collections of ink dots could produce some dark gray visual effect. In practice, the size of the dots is inversely proportional to the gray values to be represented, that is, the dots in bright image area are small, while the dots in the dark image areas are large. When the dots are small enough, and are viewed far enough away, the human eye will obtain more continuous and smooth gradation image based integration features. The normal resolution of newspaper picture is about 100 points per inch (dot per inch, DPI), and the normal resolution of book or magazine picture is about 300 DPI.

In amplitude modulation, the binary points are regularly arranged. The size of these dots changes according to the gray scale to be represented, while the dot form is not a decisive factor. For example, the laser printer simulates different shades of gray by controlling the ratio of ink coverage, while the shape of ink dots is not strictly controlled. In actual utilization of amplitude modulation technology, the output effects of binary dot pattern depend not only on the size of each point but also on the spacing of the grid related. The smaller the gap, the higher will be the resolution of the output. Granularity of the grid is limited by the printer's resolution, which is measured by dots per inch.

1.3.3.3 Frequency Modulation

In **frequency modulation** (FM) halftone technology, the size of the output black dots is fixed, but the spatial distribution of dot depends on the desired gradation (frequencies occurring within a certain area or interval between points). If the distribution is dense, darker gray will be obtained; if the distribution is sparse, lighter gray will be obtained. In other words, to obtain a darker gray effect many points (which synthesize a printing unit, also known as printing dots corresponding to one pixel in the image) should be arranged close to each other. With respect to AM halftoning technology, FM halftone technique can eliminate moles (Moiré) pattern problems due to the superposition of two or more regular pattern generated in AM halftoning techniques (Lau and Arce, 2001). The main drawback of frequency modulation technique is related to the increase of halftone dot gain. Dot gain is increments of printing unit cell size relative to the size of the original amount, which lead to reduced or compressed grayscale range of printed images, and this will reduce detail and contrast.

Recently, with the increase of printer resolution (>1,200 dpi), FM halftone technique has reached the limit. People began to study techniques combining AM and FM halftone techniques to obtain the point sets whose size and spacing are both changing as the alteration of output gradation. In other words, the size and spacing between the basic points of printing units are required to change with the required gray level. This will not only generate spatial resolution comparable to AM halftone technology but also obtain the effect of Moore pattern cancellation similar to FM halftone technique.

1.3.3.4 Halftone Output Templates

A concrete implementation of halftone output technique is first to subdivide the image output units, by combining the nearby basic binary points to form the output unit. In this way, each output unit will contain a number of basic points, if some basic binary points output black, other basic binary points output white, then the effect of different gray levels can be obtained. In other words, to output different gray levels, a set of **halftone output templates**, each corresponding to an output unit, need to be established. Each template is divided into a regular grid, and each grid point corresponds to a basic binary point. By adjusting the respective basic binary point as black or white, it is possible to make each template to output different gray levels, so as to achieve the goal of outputting grayscale images.

Example 1.4 Halftone output templates
By dividing a template into a grid of 2×2, the five different shades of gray can be output in accordance with Figure 1.10. By dividing a template into a grid of 3×3, the ten different shades of gray can be output in accordance with Figure 1.11. Because there are $C_k^n = n!/(n - k)!k!$ kinds of different ways to put k points into n units, the number of ways in these figures are not unique. Note that in both figures, if a grid is black in a certain gray level, then this grid will be still black for all gray levels that are greater than this gray level in output. ▣

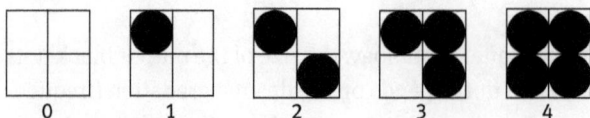

Figure 1.10: Dividing a template into a grid of 2×2 to output five gray scales.

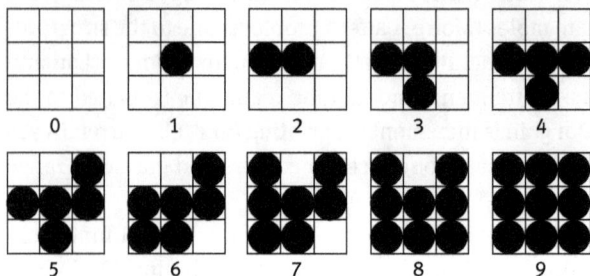

Figure 1.11: Dividing a template into a grid of 3×3 to output ten gray scales.

According to the above method of dividing the template into the grid, for the output of 256 types of gray levels, a template will need to be divided into a 16×16 units. It is seen that the visible spatial resolution of the output image will be greatly affected. Halftone output technology is only worth used under the circumstances that the output gray values of output device itself are limited. Suppose that each pixel in a 2×2 matrix may be black or white, each pixel needs one bit to represent. Taking this 2×2 matrix as a unit of halftoning output, then this unit requires 4 bits and can output five gray levels (16 kinds of modes), 0/4, 1/4, 2/4, 3/4, and 4/4 (or written 0, 1, 2, 3, 4), respectively. However, if a pixel is represented by 4 bits, this pixel may have 16 kinds of gray scales. It is seen that if the halftone output using the same number of memory cells, the smaller the output scales the larger the output size.

1.3.3.5 Dithering Technology

Halftone output technology improves the amplitude resolution of the image by reducing the spatial resolution of the image, or increases the image gray levels by sacrificing the numbers of space points. It is seen from the above discussion that if the output image with more gray scales is required, the spatial resolution of the image will be greatly reduced. On the other side, if a certain spatial resolution needs to be maintained, the output gray levels will be relatively small, or to keep the spatial details the number of gray levels cannot be too large. However, when a grayscale image has relatively small number of gray levels, the visual quality of the image will be relatively poor, such as the appearing of a false contour phenomenon (see examples and discussions in Section 2.4.2). To improve the quality of the image, dithering technique is often used, which improves the display quality of images quantized too crude, by adjusting the amplitude or variation of the image.

Dithering can be accomplished by adding a small random noise $d(x, y)$ to the original image $f(x, y)$. Since the values of $d(x, y)$ and the values of $f(x, y)$ do not have any regular connection, the false contour phenomenon caused by quantization can be reduced or eliminated.

One particular way to achieve dithering is as follows. Let b be the number of bits used for the image display, then the values of $d(x, y)$ can be obtained with a uniform probability from the following five number: $-2^{(6-b)}$, $-2^{(5-b)}$, 0, $2^{(5-b)}$, $2^{(6-b)}$. Adding the b most significant bits of a small random noise $d(x, y)$ to $f(x, y)$ will produce the final output pixel values.

Example 1.5 Dithering examples

A set of dithering examples is given in Figure 1.12. Figure 1.12(a) is a part (128×128) of an original image of 256 gray levels (see Figure 1.1(a)). Figure 1.12(b) shows the halftone output image (the same size as the original image) obtained by using the 3×3 halftone templates. Since there are only ten gray levels now, the false contour phenomenon is appearing more obvious at the face, shoulder, and other regions with relatively slow gradation changing. The original continuously variable gray scales appear now having dramatic changes with gray scale discontinuity. Figure 1.12(c) is the result of dithering for adjusting the original image, the superimposed dithering values are uniformly distributed in the range of $[-8, 8]$. Figure 1.12(d) shows the halftone output image (the same size as the original image) also obtained by using the 3×3 halftone templates, but with Figure 1.12(c) as input image. The effect of false contour phenomenon has been reduced.

The magnified partial views (42×42) of the halftone image in Figure 1.12(b) and the dither halftone image in Figure 1.12(d) are given in Figures 1.13(a) and (b), respectively. Since each pixel of the original image is represented by a matrix of 3×3 units, the image size is now 126×126 units. It is seen from the figure that the false contours become less regular due to dithering, so they are thus less likely to be observed.

As seen above, the use of dithering technique may eliminate some false contour phenomenon due to the too small number of gray levels in the image region with

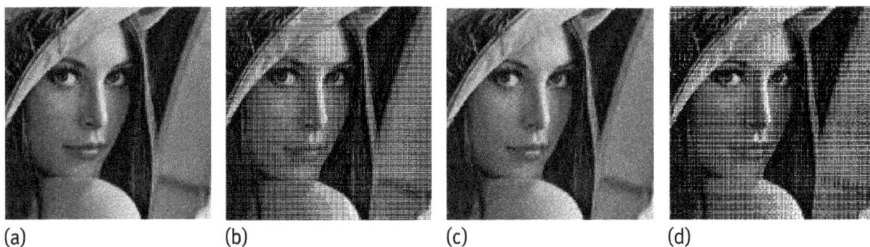

(a)　　　　　　　(b)　　　　　　　(c)　　　　　　　(d)

Figure 1.12: Dithering examples.

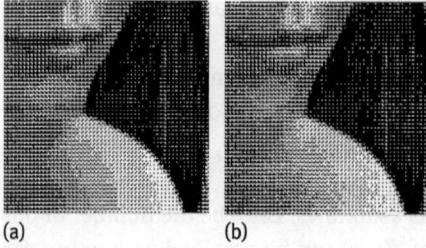

(a)   (b)

Figure 1.13: A comparison of a halftone image and a dither halftone image.

smooth gray levels. According to the principle of dithering, the larger the dithering values superimposed, the more obvious the effect of eliminating false contour. However, the superimposed dithering value also brings noise to the image, the bigger the dithering values, the greater the noise influence.   ◙

1.3.4 Image Storage

The storage of image requires the use of storage memory. The image also needs to be stored in a certain format.

1.3.4.1 Image Memory
A lot of space is required to store the image. In a computer, the smallest unit of measurement for image data is bit (b), some bigger units are bytes (1 B = 8 b), kilobytes (1 KB = 10^3 B), megabytes (1 MB = 10^6 B), gigabytes (1 GB = 10^9 B), terabytes (1 TB = 10^{12} B), etc.

Example 1.6 Image resolution and storage memory
The number of bits required to store an image is often quite large. Suppose in an image of 1024×1024, 256 gray levels is required to store, then 1 MB storage memory is needed. If this image is replaced by a color image, 3 MB storage memory is needed, which is equivalent to the memory required for storing a book of 750 pages. Video is composited by successive image frames (PAL system has 25 frames per second, while NTSC system has 30 frames per second). Suppose that the size of each frame image in color video is 640 × 480, to store 1 hour of video will require 83 G memory.   ◙

In the image processing system, large capacity and fast storage memory is essential. Commonly used **image memory** includes tape, disk, flash memory, optical disk, and magneto-optical disks and the like. The memory employed for image processing can be divided into three categories.

Fast Memory Used in Processing Computer memory is for providing fast storage function. The memory size in general microcomputer is often several gigabytes. Another memory for providing fast storage function is the special hardware card, also called a frame buffer. It is often possible to store multiple images and can be read out in video speed (25 or 30 images per second). It also allows the image zooming, vertical flip, and horizontal flip of image. The capacity of the most commonly used frame buffer is up to tens of GB.

Quick Memory for Online Recall or Online Storage   Disk is one kind of generic online storage memory; for example, commonly used Winchester disks can store hundreds of GB data. There is also a magneto-optical (MO) memory, which can store 5 GB of data on the 5¼-inch optical disk. A feature of the online storage memory is that it is often required to read data, so in general the sequential media, such as tape, are not used. For even larger storage requirements, the CD-ROM and CD-ROM tower array can be used. A CD tower can hold dozens to hundreds of CDs, some mechanical devices are required for inserting or extracting CD-ROM from the CD-ROM drive.

Infrequently Used Database (Archive) Memory   A feature of database storage requirements is very large capacity, but the reading of the data is less frequent. Commonly used write-once-read-many (WORM) optical disk can store 6 GB of data on a 12-inch disc, or 10 GB of data on a 14-inch disc. Data on WORM discs can be stored in the general environment for more than 30 years. When the main application is to read, the WORM disc can also be placed in the disc tower. A WORM optical disc tower with capacity reaching a level of TB storage can store more than one million grayscale and color images of megapixels.

1.3.4.2 Image File Format

Image file refers to a computer file containing the image data. Within the image file, in addition to the image data themselves, some general information on the description of the image, for easily to read and display an image, are also included.

In image data files, the mainly used form is raster form (also known as bitmap or pixel map), this form is consistent with people's understanding of the image (an image is a collection of image points). Its main disadvantage is that there is no direct relationship between pixels being shown, and it limits the spatial resolution of the image. The latter brings two problems: one is there will be a block effect when the image is enlarged to a certain extent; the other is that if the image is first reduced and then restored to the original size, the image will become blurred.

Image file format has a variety of different platforms, and different software systems often use diverse image file formats. Following is a brief introduction for four kinds of widely used formats.

BMP format   is a standard in Windows environment, its full name is Microsoft Device Independent (DIP) Bitmap. The image files of BMP format, also known as bitmap files, include three parts: the first is bitmap file header (also known as the header); the second is bitmap information (often called a palette); the third is bitmap array (*i. e.,* image data). One bitmap file can only store one image.

The length of bitmap file header is fixed at 54B, which gives the information on the image file type, image file size, and the starting location of the bitmap array. Bitmap information gives the length, width, number of bits per pixel (which may be 1, 4, 8, 24, corresponding respectively to monochrome, 16-color, 256-color, and true color), the compression method, the horizontal and vertical resolution information of the target device, etc. Bitmap array provides the values for each pixel in the original image (3 bytes per pixel, namely blue, green, red values), it can be stored in compressed format (only for images of 16 colors and 256 colors) or uncompressed format.

GIF format   is a standard of common image file format. It is generally 8-bit file format (one pixel uses one byte), so it can only store images up to 256 colors. The image data in GIF files are compressed image data.

The structure of GIF files is relatively complex, generally includes seven data elements: header, general palette, the image data area, and four complementary regions, wherein the header and the image data area are indispensable units.

A GIF file can store multiple images (this feature is very favorable for the animation on the web page), so the header will contain the global data for all images and the local data belongs only to the subsequent images. When there is only one image in the file, the global data and local data are consistent. When storing multiple images, each image is centralized into one image data block, the first byte of each block is an identifier, indicating the type of data block (may be an image block, extension block or the symbol marking the end of file).

TIFF format   is a format independent from operating system and file system. It is easy to exchange image data among different software. TIFF image file includes a file header (header), file directory (identification of information area), and the file directory entry (image data area). There is only a header in a file, which is located in the front end of the file. It gives the information on the order of data storage, the offset of bytes in file directory. The file directory provides the information on the number of file directory entries and includes a set of identification information that gives the address of image data area. The file directory entry is the basic unit of information storage, also known as domains. There are five kinds of domains: basic domains, information description domains, fax domains, document storage domains, and retrieval domains.

TIFF format can support image of any size. The image files can be divided into four categories: binary image, gray image, the color palette image, and full color image. A TIFF file can store multiple images or multiple copies of palette data.

JPEG Format   JPEG is a compression standard for static grayscale or color images. The space saved when using the form of lossy compression is quite large. The current digital cameras all use **JPEG format.**

JPEG standard defines only a specification for coded data stream, there is no prescribed format for image data files. Cube Microsystems company defines a JPEG File

Interchange Format (JFIF). JFIF image is a representation of grayscale or Y, C_b, C_r color components of the JPEG image. It contains a header compatible with JPEG. JFIF file usually contains a single image, the image may be a gray scale (where the data are for a single component) or colored (which includes data of Y, C_b, C_r components).

1.3.5 Image Processing

The image processing module is the central module of the image processing system, whose task of function is to complete process for image. It is also central to this book, and will be described in detail in subsequent chapters. Depending on the purpose and requirements of image processing, the process to image may take a number of different technology. The main purpose and technology for image processing include the image enhancement to improve the visual quality of the image, the restoration of degraded images to eliminate the effects of various kinds of interference, the reconstruction of scene based on different projections, the coding of image to reduce the amount of representation data for easing storage and transmission, the watermarking of image to protect the ownership of images, and the extending of these technologies to process color images, video, or multi-scale images.

　　Various kinds of image processing procedures can in general be described by the form of algorithms, and most of the available algorithms are implemented in software, so many of the image processing tasks can be completed with only the ordinary or general purpose computer. In many online applications, in order to improve processing speed or overcome the limitations of general purpose computer, the advantage of a special or dedicated hardware can be taken. In the last decade of the 20th century, various graphics cards designed to be compatible with the industry standard bus have been inserted into PC or workstation. This not only reduces costs but also promoted the development of a dedicated image processing software. Since the 21st century, the researches and applications of system on chip (SOC) and graphic processing unit (GPU) make an important progress in terms of hardware. On the other side, it also promotes and facilitates the further development of image processing software.

Example 1.7 Image resolution and the capabilities of image processing
Real-time processing of each frame in a color video of 1024×1024 pixels requires processing $1024 \times 1024 \times 8 \times 3 \times 25$ bits of data per second, corresponding to the processing speed of about 78.64 GB per second. Suppose the processing for one pixel requires ten floating-point operations (FLOPS), then for processing 1 second video it will require nearly 8 million FLOPS. Parallel computing strategy can speed up processing through the use of multiple processors to work simultaneously. The most optimistic estimates suggest that parallel computing time can be reduced to $\ln J/J$ compared to a serial computing, where J is the number of parallel processors (Bow 2002). According to this estimate, if one million parallel processors are used to handle the process of a second of video, each processor still needs to have the ability of 780,000 operations per second.　　　　　　　　　　　　　　　　　　　　　　　　□

One important fact in image processing is that special problems require special solutions. Existing image processing software and hardware can provide more and faster common tools than ever before, but to solve the specific problems further research and development of newer tools are still needed. Frequently, in order to solve a problem of image processing applications it might have different techniques, but sometimes the application looked similar require different techniques to solve. So this book is to help the reader to lay the foundation for further research and development, mainly through the introduction of the basic theory of image processing and practical techniques. At the same time, the book also introduces specific ways for solving concrete problems in image processing applications, so that the readers can take to extrapolate.

1.4 Overview of the Book

This book has eight chapters. The current chapter makes an overview of image processing. The basic concepts and terms of the image, and the expression of the image (including the image representation and display form) are introduced. A brief overview of image engineering is provided. Different modules in image processing system are discussed.

Chapter 2 is titled "Image acquisition." This chapter discusses the spatial relationship between the real world and the projected image, the brightness of world and the gray level of image, as well as the sampling and the quantization of image coordinates and gray levels. The principle of stereo imaging with several models is also introduced.

Chapter 3 is titled "Image enhancement." This chapter discusses various image operations to improve the visual quality of images, such as direct gray-level mapping techniques, histogram transforms, frequency filtering, linear spatial filtering, and nonlinear spatial filtering.

Chapter 4 is titled "Image restoration." This chapter discusses the concepts of degradation and noise, the degradation model and restoration computation, techniques for unconstrained restoration, techniques for constrained restoration, and techniques for interactive restoration. In addition, the principle of recently developed image repairing techniques is also introduced.

Chapter 5 is titled "Image reconstruction from projection." This chapter discusses various modes and principles for reconstructing image from projection, such as techniques of reconstruction by Fourier-inversion, techniques using convolution and back-projection, algebraic reconstruction techniques, and combined techniques.

Chapter 6 is titled "Image coding." This chapter discusses some fundamental concepts and principles for image coding, such as techniques based on variable length coding, techniques based on bit-plane coding, techniques based on predictive coding, and techniques using various image transforms.

Chapter 7 is titled "Image watermarking." This chapter introduces the principles and features of watermarks, as well as practical techniques for both DCT domain and DWT domain image watermarking. The watermark performance evaluation is also discussed. In addition, the relationships between watermarking and information hiding, and an image hiding method based on iterative blending are described.

Chapter 8 is titled "Color image processing." This chapter introduces the principles of color vision and provides an explanation for color chromaticity diagram. Different basic and common color models considered and their relationships are discussed. After the explanation of several basic pseudo color image enhancement techniques, the strategy for true color image processing and two groups of full color methods are provided.

Each chapter of this book is self-contained and has similar structure. After a general outline and an indication of the contents of each section in the chapter, the main subjects are introduced in several sections. In the end of every chapter, 12 exercises are provided in the section "Problems and Questions." Some of them involve conceptual understanding, some of them require formula derivation, some of them need calculation, and some of them demand practical programming. The answers or hints for both are collected in the end of the book to help readers to start.

The references cited in the book are listed at the end of the book. These references can be broadly divided into two categories. One category is related directly to the contents of the material described in this book, the reader can find from them the source of relevant definition, formula derivation, and example explanation. References of this category are generally marked at the corresponding positions in the text. The other category is to help the reader for further study, for expanding the horizons or solve specific problems in scientific research. References of this category are listed in the Section "Further Reading" at the end of each chapter, where the main contents of these references are simply pointed out to help the reader targeted to access.

For the learning of this book, some basic knowledge are generally useful:

(1) Mathematics: The linear algebra and matrix theory are important, as the image is represented by matrix and the image processing often requires matrix manipulation. In addition, the knowledge of statistics, probability theory, and stochastic modeling are also very worthwhile.

(2) Computer science: The mastery of computer software technology, the understanding of the computer architecture system, and the application of computer programming methods are very important.

(3) Electronics: Many devices involved in image processing are electronic devices, such as camera, video camera, and display screen. In addition, electronic board, FPGA, GPU, SOC, etc. are frequently used.

Some specific pre-requests for this book would be the fundamental of signal and system, and signal processing.

1.5 Problems and Questions

?

1-1 What is the meaning of each variable in continuous image $f(x, y)$ and digital image $I(r, c)$? What are their relation and differences? What are their value ranges?

1-2 Compare the advantages and disadvantages of different image display methods.

1-3 Write out the vector representation of 2-D images and 3-D images.

1-4* The baud rate is a commonly used measure of discrete data transmission. When binary is used, it is equal to the number of bits transmitted per second. It is assumed that each time one start bit is transmitted, 8 bits of information are then transmitted, and finally one stop bit is transmitted. Please provide the times required to transfer the image in the following two cases:

(1) An image of 256×256, 256 gray level is transmitted at 9,600 baud.

(2) A color image of 1024×1024, 16,777,216 color is transmitted at 38,400 baud.

1-5 Make three lists of the important characteristics of image processing, image analysis, and image understanding, respectively. What are their similarities and differences?

1-6 Look at Table 1.1 again. What other observations and conclusions can you make from it?

1-7 Give a complete list of the applications of image techniques.

1-8* If a position in a 2×2 template can represent four kinds of gray levels, then this template altogether can express how many gray levels?

1-9 Find a bitmap file, and read its head and information parts. What contents can be directly obtained from viewing the bitmap itself?

1-10 Find a GIF file for animation, and read its head and data. What are the global data and what are the local data?

1-11 Compare the heads of a TIFF file for a binary image and a TIFF file for a gray-level image, what is their difference?

1-12 List some examples to show that a 2-D image does not contain all geometric information needed to recover a real scene.

1.6 Further Reading

i

1. **Basic Concepts of Images**
   - Detailed discussions on human eyes and visual systems can be found in Aumont (1997) and Luo *et al.* (2006).

   - More explanation on the analog image (also the digital image) can be found in Pratt (2001).

   - In this book, the introduction of image techniques uses mostly gray-level image examples, though many of them can be easily extended to color images. More discussion on color images can be found in Wyszecki and Stiles (1982), Palus (1998), MarDonald and Luo (1999), Plataniotis and Venetsanopoulos (2000), and Koschan (2010).

2. **Image Engineering**

    -   The series of survey papers for image engineering can be found in Zhang (1996a, 1996b, 1997a, 1998, 1999a, 2000a, 2001a, 2002a, 2003, 2004a, 2005a, 2006a, 2007a, 2008a, 2009a, 2010, 2011a, 2012a, 2013a, 2014, 2015a, 2016, 2017). Some summaries on this survey series can be found in Zhang (1996c, 2002c, 2008b, 2015c).

    -   A comprehensive introduction to image engineering can be found in textbooks (Zhang, 2007c; 2013b).

    -   Image processing has been introduced in many textbooks; for example, Rosenfeld and Kak (1976), Pavlidis (1982), Gonzalez and Wintz (1987), Jain (1989), Gonzalez and Woods (1992, 2002, 2008), Libbey (1994), Furht *et al.* (1995), Tekalp (1995), Young *et al.* (1995), Castleman (1996), Jähne (1997), Sonka *et al.* (1999, 2008), Zhang (1999c), Pratt (2001, 2007), Russ (2002, 2016), Wang *et al.* (2002), Bovik (2005), Zhang (2012b).

    -   Image analysis has a long history, though sometimes it is blended with image processing. More information on general image analysis can be found in Serra (1982), Rosenfeld (1984), Joyce (1985), Pavlidis (1988), Young and Renswoude (1988), Young (1993), Mahdavieh and Gonzalez (1992), Huang and Stucki (1993), Beddow (1997), Lohmann (1998), Sonka *et al.* (1999, 2008), Zhang (1999c, 2005c, 2012c), ASM (2000), Kropatsch and Bischof (2001), and Gonzalez and Woods (2008).

    -   Image understanding is often discussed jointly with computer vision, and the related information can be found in Hanson and Riseman (1978), Ballard and Brown (1982), Marr (1982), Levine (1985), Horn (1986), Fu *et al.* (1987), Shirai (1987), Haralick and Shapiro (1992, 1993), Faugeras (1993), Jain *et al.* (1995), Edelman (1999), Jähne *et al.* (1999a, 1999b, 1999c), Jähne and Haußecker (2000), Sonka *et al.* (1999), Hartley and Zisserman (2000, 2004), Andreu *et al.* (2001), Ritter and Wilson (2001), Zhang (2000b, 2007d, 2012d), Shapiro and Stockman (2001), Forsyth and Ponce (2003, 2012), Snyder and Qi (2004), Szeliski (2010), Davies (2012), and Prince (2012).

3. **Image Processing System**

    -   Some practical applications and further discussions about image acquisition can be found in Ashe (2010).

    -   A detailed description of the AM halftoning technique and the FM halftoning technique, as well as the results obtained by combining the AM halftoning technique and the FM halftone technique are shown in Lau and Arce (2001). A discussion of image dithering techniques is also available in Poynton (1996).

- More information related to CID can be found in Luryi and Mastra-pasqua (2014).
- There are still many image file formats, such as Exif, PNG, JPEG 2000, and WebP. The corresponding descriptions can be easily searched.

4. **Overview of the Book**
   - The main materials in this book are extracted from the books by Zhang (2006b, 2012b).
   - More solutions for problems and questions can be found in Zhang (2002b).
   - For course teaching methods and web courses related to the content of this book, see Zhang *et al.* (2001c, 2002d), Zhang and Liu (2002e), Zhang (2004b, 2004e, 2007b, 2008c), and Zhang and Zhao (2008d, 2011b). The use of teaching courseware has a certain effect for improving teaching effectiveness, see Zhang *et al.* (1999b), Zhang (2004d, 2004e). How to make better use of images in teaching can also be found in Zhang (2005b, 2009b).

2 Image Acquisition

Image acquisition is the initial step used by computers and electronic equipment to form and treat images. Images reflect the objective world in space, time, and other properties. Therefore, image acquisition captures the spatial, temporal, and other information from a scene.

The sections of this chapter will be arranged as follows:

Section 2.1 introduces several camera imaging models (including few approximate projection models), from simple to complex and from special to general.

Section 2.2 describes some binocular imaging modes, including binocular transversal mode, binocular converging mode, and binocular axial mode.

Section 2.3 discusses some basic concepts of photometry, and then introduces a basic brightness imaging model.

Section 2.4 focuses on the sampling and quantization in image acquisition, as well as spatial resolution and amplitude resolution of images.

2.1 Spatial Relationship in Image Formation

Spatial relationship plays an important role in image acquisition. It relates and converts the spatial and temporal information in a scene to the captured image.

2.1.1 Coordinate System

In **image acquisition**, the camera is used to make a perspective projection of the 3-D scene of the objective world onto the 2-D image plane. The projection, from the space point of view, can be described by imaging transformation (also known as geometrical perspective transformation or perspective transformation). Imaging transformation involves the conversion between different spatial coordinate systems. In consideration that the end result of the image acquisition is to obtain digital images being input into the computer, the imaging coordinate systems involved in imaging scene of 3-D space include the following four coordinate systems.

2.1.1.1 World Coordinate System
World coordinate system is also known as the real-world coordinate system XYZ, which represents the absolute coordinates of the objective world (it is also called objective coordinate system). General 3-D scenes are using this coordinate system for representation.

2.1.1.2 Camera Coordinate System
Camera coordinate system is the coordinate system xyz, which is developed by taking the camera as center (origin) and the camera optical axis as z axis.

DOI 10.1515/9783110524116-002

2.1.1.3 Image Plane Coordinate System

Image plane coordinate system is the coordinate system $x'y'$ at the imaging plane of camera. Generally, the xy plane is taken as parallel to the plane of the camera coordinate system, as well as the x-axis and the x'-axis, y-axis and the y'-axis coincides, respectively, so that the origin of the image plane is on the optical axis of the camera.

2.1.1.4 Computer Image Coordinate System

Computer image coordinate system is the coordinate system MN that is inside the computer for representing image. The final image is stored by a memory within the computer, so the projection coordinates on image plane needs to be converted to computer image coordinate system.

According to the above mutual relations of several different coordinate systems, several different (camera) imaging model can be formed. In the following, the step-by-step descriptions from simple to complex are provided.

2.1.2 Perspective Transformations

Image acquisition includes the projection of a 3-D scene onto a 2-D image in which the perspective projection/transformation plays an important role.

2.1.2.1 Projection Model

A model of image capturing by a camera with **perspective projection** is shown in Figure 2.1. Three coordinate systems are involved: The world coordinate system (X, Y, Z), the camera coordinate system (x, y, z), and the image coordinate system (x', y'). Here, suppose the camera coordinate system is aligned with the world coordinate system, and the image coordinate system is aligned with the camera coordinate system. This camera position is known as the normal position. The focal

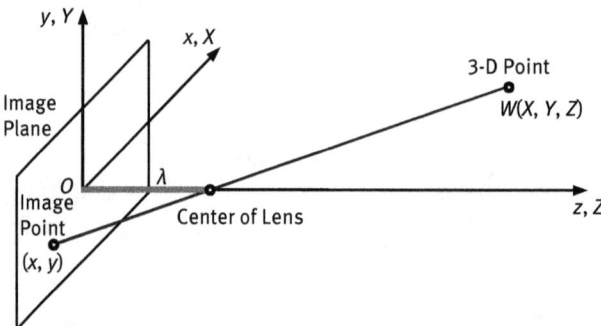

Figure 2.1: A basic image capturing model.

length of the camera lens is λ. A 3-D point $W(X, Y, Z)$ is projected onto the point $(x, y) = (x', y')$ at the image plane.

According to the properties of similar triangles, the relationship between a 3-D point in the world space and its 2-D projection at the image plane is established by

$$\frac{x}{\lambda} = -\frac{X}{Z - \lambda} = \frac{X}{\lambda - Z} \tag{2.1}$$

$$\frac{y}{\lambda} = -\frac{Y}{Z - \lambda} = \frac{Y}{\lambda - Z} \tag{2.2}$$

where the negative signs in front of X and Y indicate the inversion of the image point with respect to the space point. The coordinates of the image point are obtained by

$$x = \frac{\lambda X}{\lambda - Z} \tag{2.3}$$

$$y = \frac{\lambda Y}{\lambda - Z} \tag{2.4}$$

The above projection maps a line in 3-D space onto a line at the image plane (with the exception that the line to be projected is perpendicular to the image plane). A rectangle in 3-D space could be any quadrangle (determined by four points) at the image plane, after the projection. Therefore, such a projection is referred to as a four-point mapping. The transformations in eq. (2.3) and eq. (2.4) are nonlinear as the coordinate values are involved in the division.

Example 2.1 Telecentric imaging and supercentric imaging
In general standard optical imaging system, the light beams are converged. It has a significant adverse effect for the optical measurement; see Figure 2.2(a). If the position of the object is changing, the image becomes larger when the object is close to the lens and becomes smaller when the object is away from the lens. The depth of the object could not be obtained from the image, unless the object is placed on the known distance, otherwise the measurement error is inevitable.

If the **aperture position** is moved to the convergence point of parallel light (F_2) a telecentric imaging system can be obtained, as shown in Figure 2.2(b). In this case, the main ray (the light beam passing through the center of aperture) is parallel to the optical axis in the object space and the small change of object position would not change the size of the object image. Of course, the farther away the object is from the focus position, the stronger the object is blurred. However, the center of blur disk does not change the position. The disadvantage of telecentric imaging is that the diameter of telecentric lens should reach at least the size of the object to be imaged. Thus, using a telecentric imaging is very expensive for very big object.

If the aperture is put even closer to the image plane than the convergence point of parallel light beams, as in Figure 2.2(c), the main ray becomes the converging lines of space. In contrast to the standard imaging of Figure 2.2(a), the farther the object, the

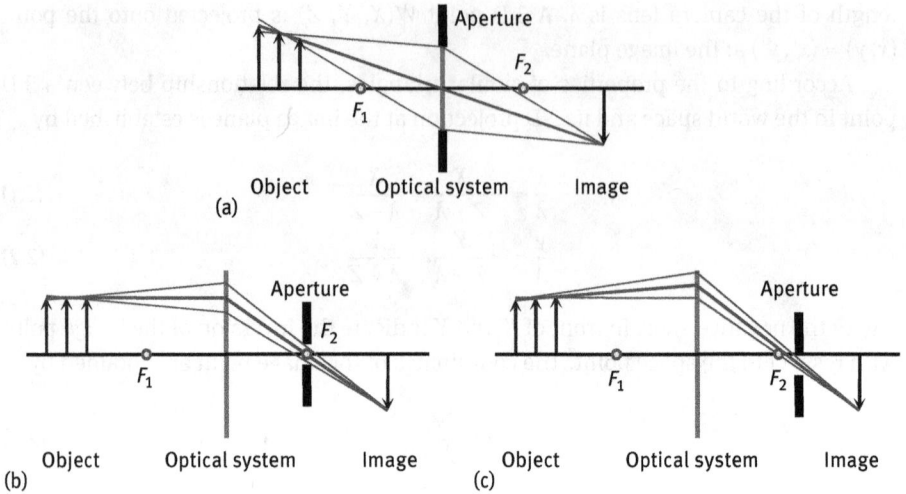

Figure 2.2: Moving the aperture position can change the property of the optical system.

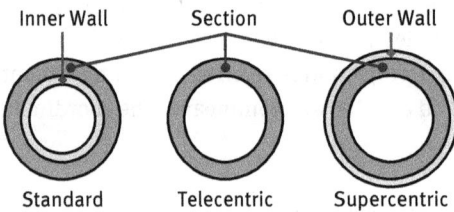

Figure 2.3: Comparison of three imaging ways.

bigger the object looks. This imaging technique is called supercentric imaging. One of its characteristics is that the surface parallel with the optical axis can be seen. In Figure 2.3, three imaging techniques for viewing a thin-walled cylinder along the optical axis are illustrated. The standard imaging can see the cross section and the inner wall, the telecentric imaging can only see the cross section, while the supercentric imaging can see both the cross section and an outer wall (Jähne, 2004).

2.1.2.2 Homogeneous Coordinates

To express these transformations in the form of a linear matrix, which is more convenient for manipulation, the homogeneous coordinates are often used.

The **homogeneous coordinates** of a point in an XYZ coordinate system are defined as (kX, kY, kZ, k), where k is an arbitrary but nonzero constant. For example, a point in an XYZ coordinate system can be expressed as a vector:

$$W = \begin{bmatrix} X\ Y\ Z \end{bmatrix}^{\mathrm{T}} \tag{2.5}$$

The vector in the homogeneous coordinates is

$$W_h = \begin{bmatrix} kX & kY & kZ & k \end{bmatrix}^T \tag{2.6}$$

By defining the matrix of perspective projection as

$$P = \begin{bmatrix} 1 & 0 & 0 & 0 \\ 0 & 1 & 0 & 0 \\ 0 & 0 & 1 & 0 \\ 0 & 0 & -1/\lambda & 1 \end{bmatrix} \tag{2.7}$$

The perspective projection of W_h is given by

$$c_h = PW_h = \begin{bmatrix} 1 & 0 & 0 & 0 \\ 0 & 1 & 0 & 0 \\ 0 & 0 & 1 & 0 \\ 0 & 0 & -1/\lambda & 1 \end{bmatrix} \begin{bmatrix} kX \\ kY \\ kZ \\ k \end{bmatrix} = \begin{bmatrix} kX \\ kY \\ kZ \\ -kZ/\lambda + k \end{bmatrix} \tag{2.8}$$

The elements of c_h are the camera coordinates in the homogeneous coordinates. They can be converted to Cartesian coordinates by dividing the first three elements of c_h by the fourth one

$$c = \begin{bmatrix} x & y & z \end{bmatrix}^T = \begin{bmatrix} \dfrac{\lambda X}{\lambda - Z} & \dfrac{\lambda Y}{\lambda - Z} & \dfrac{\lambda Z}{\lambda - Z} \end{bmatrix}^T \tag{2.9}$$

The first two elements of c are just the image coordinates of a 3-D space point (X, Y, Z) after projection.

2.1.2.3 Inverse Projection
Inverse projection is used to determine the coordinates of a 3-D point according to its projection on a 2-D image. Following eq. (2.8), this projection can be expressed by

$$W_h = P^{-1}c_h \tag{2.10}$$

where the inverse projection matrix P^{-1} is

$$P^{-1} = \begin{bmatrix} 1 & 0 & 0 & 0 \\ 0 & 1 & 0 & 0 \\ 0 & 0 & 1 & 0 \\ 0 & 0 & 1/\lambda & 1 \end{bmatrix} \tag{2.11}$$

Now consider a 2-D image point $(x', y', 0)$, which can be represented by a homogeneous form as

$$\mathbf{c}_h = \begin{bmatrix} kx' & ky' & 0 & k \end{bmatrix}^T \qquad (2.12)$$

By integrating eq. (2.12) into eq. (2.10), the coordinates of a 3-D point in homogeneous form are

$$\mathbf{W}_h = \begin{bmatrix} kx' & ky' & 0 & k \end{bmatrix}^T \qquad (2.13)$$

The corresponding coordinates in a Cartesian system are

$$\mathbf{W} = \begin{bmatrix} X & Y & Z \end{bmatrix}^T = \begin{bmatrix} x' & y' & 0 \end{bmatrix}^T \qquad (2.14)$$

It is clear that the Z coordinate value of a 3-D point could not be uniquely determined by the image point (x', y'), as Z is always zero for any 3-D point. In other words, the image point (x', y') now corresponds to all points lying on the line that pass through $(x', y', 0)$ and $(0, 0, *)$, as shown in Figure 2.1.

Solving X and Y from eq. (2.3) and eq. (2.4) yields

$$X = \frac{x'}{\lambda}(\lambda - Z) \qquad (2.15)$$

$$Y = \frac{y'}{\lambda}(\lambda - Z) \qquad (2.16)$$

It is clear that recovering a 3-D point from its projection on an image by inverse perspective transformation is impossible unless its Z coordinate value is known.

2.1.3 Approximate Projection Modes

Perspective projection is a precise projection mode (with nonlinear mapping), but the calculation and analysis are relatively complex. In order to simplify the operation, some (linear) approximate projection modes can be used when the object distance is much larger than the scale of the object itself.

2.1.3.1 Orthogonal Projection
In the **orthogonal projection**, the Z-coordinate of the 3-D spatial point is not considered (the information for object distance are lost), which is equivalent to mapping the 3-D spatial point directly/vertically to the image plane in the direction of the optical axis of the camera. Figure 2.4 shows a bar-shaped object projected onto the Y-axis (corresponding to the YZ profile in Figure 2.1) in both orthogonal projection and perspective projection.

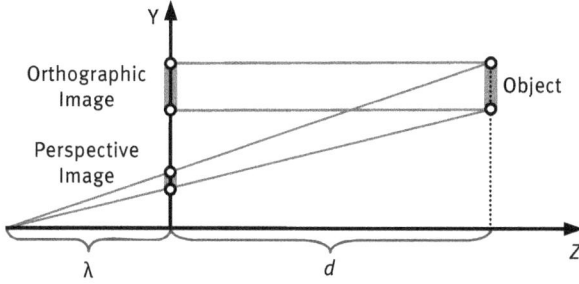

Figure 2.4: Comparison of orthographic projection and perspective projection.

The result of orthographic projection shows the true scale of the cross section of the object, while the result of the perspective projection is related to the object distance d. The orthographic projection can be seen as the particular perspective projection in which the focal length λ is infinite, so the projection transformation matrix can be written as

$$P = \begin{bmatrix} 1 & 0 & 0 & 0 \\ 0 & 1 & 0 & 0 \\ 0 & 0 & 1 & 0 \\ 0 & 0 & 0 & 1 \end{bmatrix} \qquad (2.17)$$

2.1.3.2 Weak Perspective Projection

Weak perspective projection (WPP) is an approximation to the perspective projection. In the weak perspective projection, the first step is to make an orthogonal projection to the plane where the mass center of object is located and is parallel to the image plane; the second step is to make a perspective projection to the image plane. The perspective projection of step 2 here can be achieved by means of equal scaling within the image plane. Let the perspective scaling factor be s, then

$$s = \lambda/d \qquad (2.18)$$

is the ration of the focal length over the object distance. By considering s, the projection matrix of the weak perspective can be written as

$$P = \begin{bmatrix} s & 0 & 0 & 0 \\ 0 & s & 0 & 0 \\ 0 & 0 & 1 & 0 \\ 0 & 0 & 0 & 1 \end{bmatrix} \qquad (2.19)$$

The orthogonal projection in the weak perspective projection does not consider the Z-coordinate of the 3-D space point but also changes the relative position of each point in the projection plane (corresponding to the change of size). This can make a greater impact on object points that have relatively large distance with the optical axis of the camera. It can be proved that the error on image is the first-order infinitesimal of the error on object.

2.1.3.3 Paraperspective Projection

Paraperspective projection is also a projection method located between the ortho-gonal projection and perspective projection (Dean *et al.*, 1995). A schematic of the paraperspective projection is shown in Figure 2.5. In the figure, the world coordinate system coincides with the camera coordinate system, and the focal length of the cam-era is λ. The image plane intersects the Z-axis perpendicularly at point $(0, 0, \lambda)$. Point C is the center of mass of the object S, and the distance from the origin in Z direction is d.

Given a projection plane at $Z = d$ and parallel to the image plane, the process of paraperspective projection has two steps:

(1) Given a particular object S (one of the objects in scene), projecting it parallelly to the projection plane, where each projection line is parallel to the line OC (not necessarily perpendicular to the projection plane).

(2) The above projection result on the projection plane is then projected onto the image plane. The projection result on the image plane will be reduced to λ/d (the same as in the weak perspective projection) because the projection plane and the image plane are parallel.

The first step described above takes into account the effects of the pose and perspect-ive, which preserves the relative position of the object points in the projection plane. The second step takes into account the effects of distance and other positions. It can be shown that the image error is the second-order infinitesimal of the object error, so the parallel perspective projection is closer to the perspective projection than the weak perspective projection.

Example 2.2 Comparison of projection modes
The perspective projection mode and its three approximate projection modes are shown together in Figure 2.6 for comparison. ◻

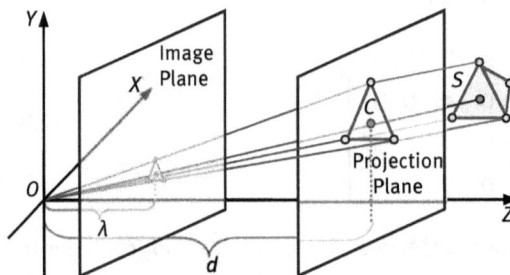

Figure 2.5: Paraperspective projection imaging diagram.

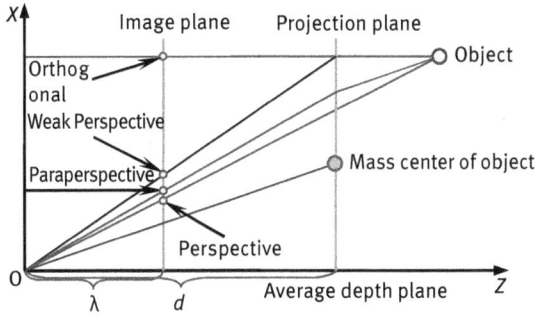

Figure 2.6: Different projection modes.

2.1.4 A General Camera Model

In image engineering, image acquisition is accomplished with the help of certain equipment that can capture the radiation and convert it to a signal. A digital camera is one such piece of equipment. A general camera model is described in the following. In addition, the formulas for perspective projection are presented.

2.1.4.1 Model Conversion

In the above discussion, the camera coordinate system is supposed to be aligned with the world coordinate system. In a real situation, this assumption is not always applicable. A more general case is depicted in Figure 2.7. In Figure 2.7, the camera coordinate system is separated from the world coordinate system, but the image coordinate system is still aligned with the camera coordinate system. The offset of the center of the camera (the origin of the camera coordinate system) from the origin of the world coordinate system is denoted $D = [D_x, D_y, D_z]$. The camera has been panned by an angle γ (the angle between x and X) and tilted by an angle α (the angle between z and Z).

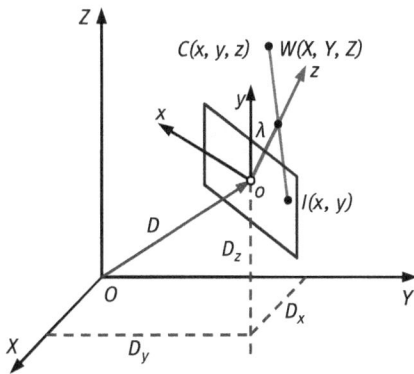

Figure 2.7: A general image-capturing model.

The above **general camera model** can be generated from the basic model shown in Figure 2.1, using the following consecutive steps:
(1) Displace the camera center from the origin of the world coordinate system.
(2) Pan the x-axis by an angle γ (around the z-axis).
(3) Tilt the z-axis by an angle α (around the x-axis).

Moving the camera with respect to the world coordinate system is equivalent to inversely moving the world coordinate system with respect to the camera. The first step of the displacement of the origin of the world coordinate system to the camera center can be accomplished by the following translation transformation matrix

$$
T = \begin{bmatrix} 1 & 0 & 0 & -D_x \\ 0 & 1 & 0 & -D_y \\ 0 & 0 & 1 & -D_z \\ 0 & 0 & 0 & 1 \end{bmatrix}
\tag{2.20}
$$

In other words, a point (D_x, D_y, D_z) whose homogeneous form is D_h will be located at the origin of the world coordinate system after the transformation of TD_h.

Further, consider the coincident of coordinate systems. The pan angle γ is the angle between the x- and X-axes, which are parallel in original situation. The rotation about the x-axis by the angle γ can be achieved by rotating the camera around the z-axis with the angle γ counterclockwise, where the rotation transformation matrix is

$$
R_\gamma = \begin{bmatrix} \cos\gamma & \sin\gamma & 0 & 0 \\ -\sin\gamma & \cos\gamma & 0 & 0 \\ 0 & 0 & 1 & 0 \\ 0 & 0 & 0 & 1 \end{bmatrix}
\tag{2.21}
$$

Similarly, the tilt angle α is the angle between the z- and Z-axes. The rotation about z-axis by the angle α can be achieved by rotating the camera around the x-axis with the angle α counterclockwise, where the rotation transformation matrix is

$$
R_\alpha = \begin{bmatrix} 1 & 0 & 0 & 0 \\ 0 & \cos\alpha & \sin\alpha & 0 \\ 0 & -\sin\alpha & \cos\alpha & 0 \\ 0 & 0 & 0 & 1 \end{bmatrix}
\tag{2.22}
$$

The above two rotation matrices can be cascaded into a single matrix

$$
R = R_\alpha R_\gamma = \begin{bmatrix} \cos\gamma & \sin\gamma & 0 & 0 \\ -\sin\gamma\cos\alpha & \cos\alpha\cos\gamma & \sin\alpha & 0 \\ \sin\alpha\sin\gamma & -\sin\alpha\cos\gamma & \cos\alpha & 0 \\ 0 & 0 & 0 & 1 \end{bmatrix}
\tag{2.23}
$$

2.1.4.2 Perspective Projection

According to the above discussion, a 3-D world point in its homogeneous form W_h after **perspective projection** becomes

$$c_h = PRTW_h \tag{2.24}$$

where P is defined in eq. (2.7).

By dividing the first two elements of c_h by its fourth element, the Cartesian coordinates (x, y) for a 3-D world point $W(X, Y, Z)$ can be obtained by

$$x = \lambda \frac{(X - D_x)\cos\gamma + (Y - D_y)\sin\gamma}{-(X - D_x)\sin\alpha\sin\gamma + (Y - D_y)\sin\alpha\cos\gamma - (Z - D_z)\cos\alpha + \lambda} \tag{2.25}$$

$$y = \lambda \frac{-(X - D_x)\sin\gamma\cos\alpha + (Y - D_y)\cos\alpha\cos\gamma + (Z - D_z)\sin\alpha}{-(X - D_x)\sin\alpha\sin\gamma + (Y - D_y)\sin\alpha\cos\gamma - (Z - D_z)\cos\alpha + \lambda} \tag{2.26}$$

Example 2.3 Computing image-plane coordinates

Suppose a camera is mounted in a position for surveillance as shown in Figure 2.8. The center of the camera is located at $(0, 0, 1)$, the focal length of the camera is 0.05 m, the pan angle is 135°, and the tilt angle is also 135°. The task is to determine the image coordinates for space point $W(1, 1, 0)$.

According to the above discussion, it has to find the procedure that moves the camera from the normal position in Figure 2.1 to the particular position in Figure 2.8. The three required steps are depicted in Figure 2.9. Figure 2.9(a) shows the normal position of the camera. The first step is to displace the camera out of the origin, as shown in Figure 2.9(b). The second step is to pan the camera by the angle γ around the z-axis, as shown in Figure 2.9(c). The third step is to rotate the camera around the x-axis and tilt it by the angle α relative to the z-axis, as shown in Figure 2.9(d).

Putting all parameters into eq. (2.22) and eq. (2.23), the image coordinate values of the 3-D point $W(1, 1, 0)$ are $x = 0$ m and $y = -0.008837488$ m. ▢

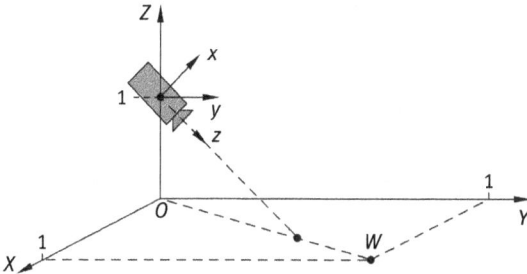

Figure 2.8: Computing image-plane coordinates.

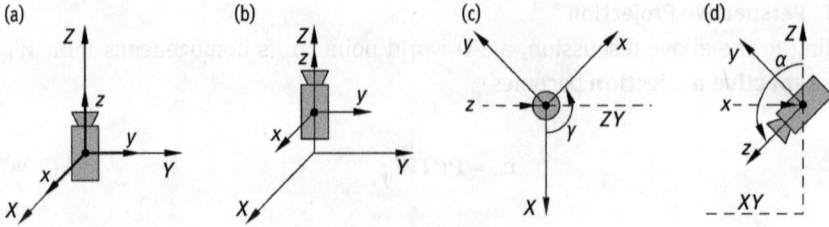

Figure 2.9: Steps used to bring camera from a normal position to the required position.

2.1.5 Common Imaging Model

In the imaging model that is more general than the abovementioned general camera model, there are two factors to consider in addition to the noncoincidence of the world coordinate system: the camera coordinate system and the image plane coordinate system. First, the camera lens will be distorted in practice, so the imaging position on the image plane will be shifted from the results of perspective projection calculated with the aforementioned formula. Second, the unit of image coordinates used in the computer is the number of discrete pixels in the memory, so the coordinates on the image plane (here, the continuous coordinates are still used) need to be converted to integer. Figure 2.10 shows a very **common imaging model** with all these factors in mind.

The imaging transformation from the objective scene to the digital image can be viewed by the following four steps (see Figure 2.11) (Tsai, 1987):

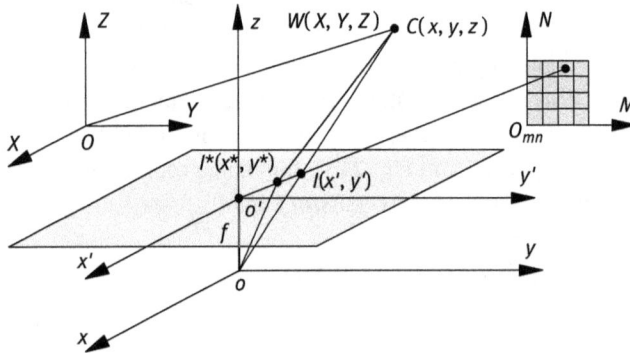

Figure 2.10: Common imaging model.

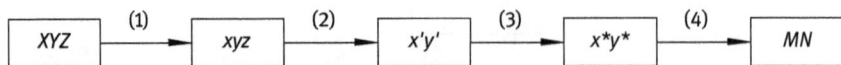

Figure 2.11: Imaging transformation from the objective scene to the digital image.

(1) Transformation from world coordinates (X, Y, Z) to camera 3-D coordinates (x, y, z). Consider the case of rigid bodies, the transformation can be expressed as

$$\begin{bmatrix} x \\ y \\ z \end{bmatrix} = R \begin{bmatrix} X \\ Y \\ Z \end{bmatrix} + T \tag{2.27}$$

where R and T are a 3×3 rotation matrix (actually a function of the angles between the axes of the three coordinate axes of the two coordinate systems) and a 1×3 translation vector, respectively:

$$R \equiv \begin{bmatrix} r_1 & r_2 & r_3 \\ r_4 & r_5 & r_6 \\ r_7 & r_8 & r_9 \end{bmatrix} \tag{2.28}$$

$$T \equiv \begin{bmatrix} T_X & T_y & T_z \end{bmatrix}^{\mathrm{T}} \tag{2.29}$$

(2) Transformation from the camera 3-D coordinates (x, y, z) to the distortion-free image plane coordinates (x', y'):

$$x' = \lambda \frac{X}{z} \tag{2.30}$$

$$y' = \lambda \frac{y}{z} \tag{2.31}$$

(3) Transformation from the distortion-free image plane coordinate (x', y') to the shifted image plane coordinate (x^*, y^*) caused by the lens radial distortion:

$$x^* = x' - R_x \tag{2.32}$$

$$y^* = y' - R_y \tag{2.33}$$

where R_x and R_y represent the radial distortion of the lens. Most of the lens have a certain degree of radial distortion. Although this distortion in general has no large impact on the human eye, it should be corrected in the optical measurement to avoid a greater error. Theoretically, the lens will have two types of distortions, that is, radial distortion and tangential distortion. Since tangential distortion is relatively small, the radial distortion is only to be considered in general industrial machine vision applications (Tsai, 1987). Radial distortion can be expressed as

$$R_x = x^* (k_1 r^2 + k_2 r^4 + \cdots) \approx x^* k r^2 \tag{2.34}$$

$$R_y = x^* (k_1 r^2 + k_2 r^4 + \cdots) \approx y^* k r^2 \tag{2.35}$$

where

$$r = \sqrt{x^{*2} + y^{*2}} \tag{2.36}$$

In eq. (2.34) and eq. (2.35), $k = k_1$. The reason for this approximation simplification is that the higher order terms of r in practice are negligible. Therefore, k_2
is not considered. According to the principle of physics, the radial distortion of
a point in the image is proportional to the distance from this point to the optical
axis of the lens (Shapiro and Stockman, 2001).

(4) Transformation from the actual image plane coordinates (x^*, y^*) to the computer
image coordinates (M, N) is

$$M = \mu \frac{x^* M_x}{S_x L_x} + O_m \tag{2.37}$$

$$N = \frac{y^*}{S_y} + O_n \tag{2.38}$$

where, M and N are the numbers of row pixels and column pixels in the computer memory (computer coordinates), respectively; O_m and O_n are the number
of rows and columns for the central pixel of the computer memory; S_x is the distance between two adjacent sensors in the x direction (scan line direction), S_y is
the distance between two adjacent sensors in the y direction; L_x is the number of
sensor elements in the X direction; M_x is the number of samples in a row (number
of pixels) of the computer. μ in eq. (2.37) is an uncertainty of image scaling factor
depending on the camera. When a CCD camera is used, the image is progressively scanned. The distance between adjacent pixels in the y' direction is also
the distance between the adjacent CCD sensor elements. However, in the x' direction, certain uncertainties will be introduced due to the time difference between
the image acquisition hardware and the camera scanning hardware, or the scanning time inaccuracies of camera itself. These uncertainties can be described by
introducing uncertainty of image scaling factors.

By combining the last three steps, it is possible to obtain an equation that associates
the computer image coordinates (M, N) with the 3-D coordinates (x, y, z) of the object
points in the camera system:

$$\lambda \frac{x}{z} = x' = x^* + R_x = x^*(1 + kr^2) = \frac{(M - O_m)S_x L_x}{\mu M_x}(1 + kr^2) \tag{2.39}$$

$$\lambda \frac{y}{z} = y' = y^* + R_y = y^*(1 + kr^2) = (N - O_n)S_y(1 + kr^2) \tag{2.40}$$

Taking eq. (2.28) and eq. (2.29) into the above two equations, we get

$$M = \lambda \frac{r_1 X + r_2 Y + r_3 Z + T_x}{r_7 X + r_8 Y + r_9 Z + T_z} \frac{\mu M_x}{(1 + kr^2)S_x L_x} + O_m \tag{2.41}$$

$$N = \lambda \frac{r_4 X + r_5 Y + r_6 Z + T_y}{r_7 X + r_8 Y + r_9 Z + T_z} \frac{1}{(1 + kr^2)S_x} + O_n \tag{2.42}$$

2.2 Stereo Imaging

Stereo imaging techniques are effective for capturing the depth information that is often lost in normal image acquisition processes, because the projection from 3-D to 2-D causes all the world points along the projection line to be converted to an image point. In stereo imaging, two separate images of a scene are captured. These two images can be obtained by using two cameras located in two locations simultaneously or by using one camera that moves from one location to another and takes two images consecutively.

There are several models for stereo imaging. First, a basic model will be described, while its variations will be introduced later.

2.2.1 Parallel Horizontal Model

The parallel horizontal model is the basic model used in stereo imaging. Several of its variations are now popularly employed in different applications.

2.2.1.1 Disparity and Depth

The **parallel horizontal model** is depicted in Figure 2.12. Two cameras are located along a horizontal line. Their optical axes are parallel. The coordinate systems of these two cameras are only different by the positions of their origins (*i. e.*, all their corresponding axes are parallel). The distance between their origins is called the baseline of these systems.

For a world point W viewed by two cameras, its projections on two images are located differently in two coordinate systems. Such a difference is called a disparity, and it is this difference that comprises the depth information.

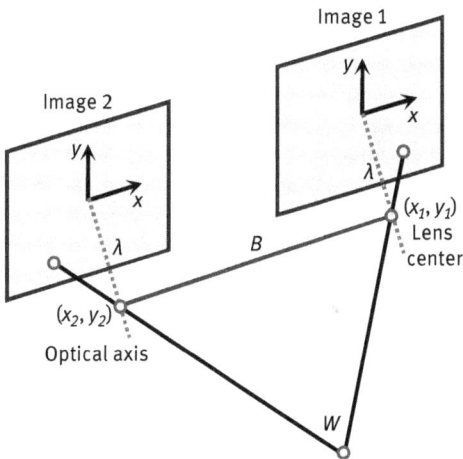

Figure 2.12: Parallel horizontal model.

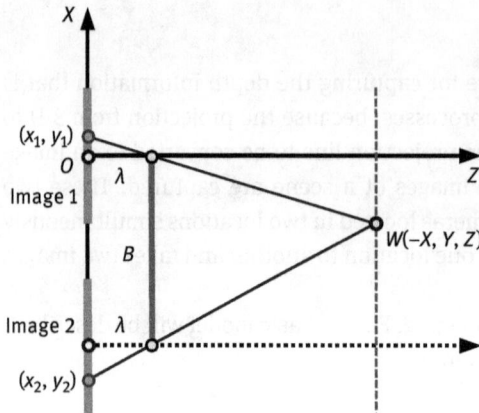

Figure 2.13: Disparity in the parallel horizontal model.

Taking a profile (corresponding to the XZ plane) of Figure 2.12 and putting the first camera coordinate system coincident with the world coordinate system, the parallel horizontal model can be depicted by Figure 2.13.

In Figure 2.12, the X coordinate of a world point W under consideration is negative. Following eq. (2.1), one equation can be obtained from the first camera coordinate system as

$$X = \frac{x_1}{\lambda}(Z - \lambda) \tag{2.43}$$

Similarly, another equation can be obtained from the second camera coordinate system as

$$B - X = \frac{-(x_2 + B)}{\lambda}(Z - \lambda) \tag{2.44}$$

Taking both eq. (2.43) and eq. (2.44) and solving them for Z gives

$$\frac{B\lambda}{Z - \lambda} = x_1 - x_2 + B \tag{2.45}$$

In eq. (2.45), the absolute value of the right side is just the disparity. Suppose this absolute value is represented by d. The depth Z can be obtained by

$$Z = \lambda\left(1 + \frac{B}{d}\right) \tag{2.46}$$

In eq. (2.46), both B and λ are determined by the characteristics (position and focus) of the camera system. Once the disparity between the two image points corresponding to the same world point is determined, the distance between the world point and the camera can be easily calculated. With the Z value in hand, the X and Y coordinates of the world point can also be determined.

2.2.1.2 Angular-Scanning Model

In the **angular-scanning model**, the stereo system rotates to capture a panorama image composed of various views. The camera used is called an angle-scanning camera. The principle of such a system can be explained with the help of Figure 2.14. When using an angle-scanning camera to capture images, the pixels are uniformly distributed in the image plane according to the azimuth and elevation of the lens. If the XZ plane is considered the equator plane of the earth and the Y-axis is pointing to the North Pole, the azimuth will correspond to the longitude and the elevation will correspond to the latitude. In Figure 2.14, the azimuth is the angle between the YZ plane and the line connecting the camera center C and the world point W, and the elevation is the angle between the XZ plane and the plane including the X-axis and the world point W.

The distance between the world point and its image point can be represented with the help of the azimuth. According to Figure 2.13, the following two relations can be established

$$\tan\theta_1 = \frac{X}{Z} \tag{2.47}$$

$$\tan\theta_2 = \frac{B-X}{Z} \tag{2.48}$$

Combining eq. (2.47) and eq. (2.48) and solving it for Z of point W yields

$$Z = \frac{B}{\tan\theta_1 + \tan\theta_2} \tag{2.49}$$

Equation (2.49) establishes the relation between the distance Z (between a world point and an image point, i. e., the depth information) and the tangents of the two azimuths. Equation (2.49) can also be written in a form like eq. (2.46), because the influences of the disparity and the focus length are all implicitly included in the azimuth. Finally,

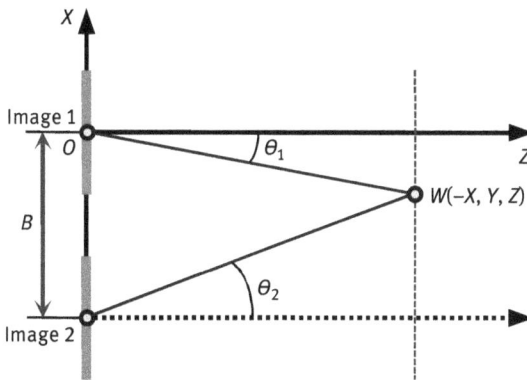

Figure 2.14: Stereoscopic imaging by an angular-scanning camera.

if the elevation is ϕ, which is common for two cameras, the X and Y coordinates of point W are

$$X = Z \tan \theta_1 \tag{2.50}$$

$$Y = Z \tan \phi \tag{2.51}$$

2.2.2 Focused Horizontal Model

The two axes of two cameras are not necessarily parallel. They can also be focused at some points. Such a group of models is called the **focused horizontal model**. Only the case shown in Figure 2.15 is considered as an example. In Figure 2.15, the arrangement in the XZ plane is depicted. Such a camera system rotates one camera coordinate system clockwise and rotates another camera coordinate system counterclockwise. In Figure 2.15, the baseline is still B. Two optical axes are crossed at point $(0, 0, Z)$ with an unknown angle 2θ.

Now consider how to determine the coordinates of a world point $W(X, Y, Z)$, once the image coordinates (x_1, y_1) and (x_2, y_2) are known. From the triangle enclosed by the X and Z axes as well as the connecting line between the focus points of the two cameras, it is easy to obtain

$$Z = \frac{B \cos \theta}{2 \sin \theta} + \lambda \cos \theta \tag{2.52}$$

As shown in Figure 2.15, two perpendicular lines from point W to the two optical axes of the cameras can be drawn, respectively. Since the two angles between these two perpendicular lines to the X-axis are all θ,

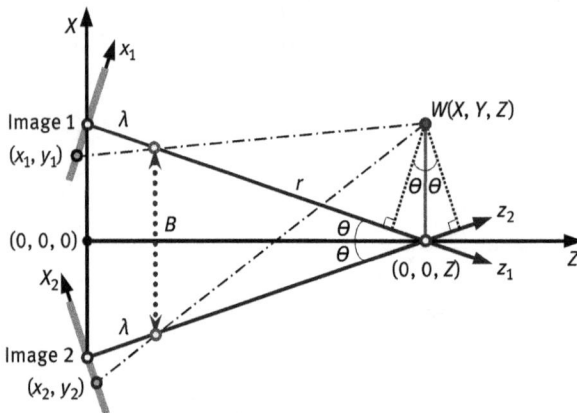

Figure 2.15: Disparity in the focused horizontal model.

$$\frac{x_1}{\lambda} = \frac{X \cos \theta}{r - X \sin \theta} \tag{2.53}$$

$$\frac{x_2}{\lambda} = \frac{X \cos \theta}{r + X \sin \theta} \tag{2.54}$$

where r is the unknown distance from any camera center to the focus point of the two cameras. Taking both eq. (2.53) and eq. (2.54) and discarding r and X gives

$$\frac{x_1}{\lambda \cos \theta + x_1 \sin \theta} = \frac{x_2}{\lambda \cos \theta - x_2 \sin \theta} \tag{2.55}$$

Substituting eq. (2.55) into eq. (2.52), eq. (2.52) becomes

$$Z = \frac{B \cos \theta}{2 \sin \theta} + \frac{2 x_1 x_2 \sin \theta}{d} \tag{2.56}$$

Similar to eq. (2.46), eq. (2.56) relates also the distance Z between the world point and the image plane with a disparity d. However, to solve eq. (2.33), only $d(x_1 - x_2)$ is needed, while to solve eq. (2.56), the values of x_1 and x_2 are also needed. On the other hand, it is seen from Figure 2.15 that

$$r = \frac{B}{2 \sin \theta} \tag{2.57}$$

Substituting eq. (2.44) into eq. (2.40) or eq. (2.41) gives

$$X = \frac{B}{2 \sin \theta} \frac{x_1}{\lambda \cos \theta + x_1 \sin \theta} = \frac{B}{2 \sin \theta} \frac{x_2}{\lambda \cos \theta - x_2 \sin \theta}. \tag{2.58}$$

2.2.3 Axis Model

In the **axis model**, two cameras are arranged along the optical axis. One camera captures the second image after the first image by moving along the optical axis. The second image can be obtained at a point closer or further away from the world point than the first image, as shown in Figure 2.16. In Figure 2.16, only the XZ plane is depicted. The first camera coordinate system is coincident with the world coordinate system and the second camera moves toward the world point. Both image coordinate systems are coincident with their corresponding camera coordinate systems. The only difference between the two camera coordinate systems is the difference along the Z-axis, ΔZ.

For each camera, eq. (2.1) can be applied. This gives (only the coordinate X is considered here, as the computation for Y would be similar)

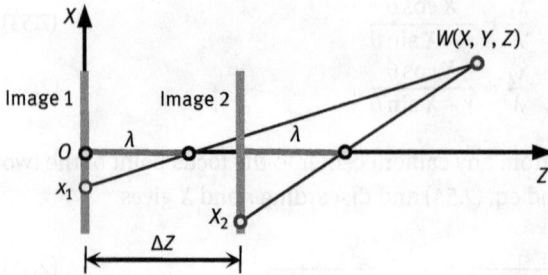

Figure 2.16: Disparity in the axis model.

$$\frac{X}{-x_1} = \frac{Z - \lambda}{\lambda} \qquad (2.59)$$

$$\frac{X}{-x_2} = \frac{Z - \lambda - \Delta Z}{\lambda} \qquad (2.60)$$

Combining eq. (2.59) and eq. (2.60) gives

$$X = \frac{\Delta Z}{\lambda} \frac{x_1 x_2}{x_1 - x_2} \qquad (2.61)$$

$$Z = \lambda + \frac{\Delta Z x_2}{x_2 - x_1} \qquad (2.62)$$

Compared to the parallel horizontal model, the common viewing field of the two cameras in the axis model is just the viewing field of the second camera (near the world point). Therefore, the boundary of the common viewing field can be easily determined. In addition, since the camera moves along the optical axis, the occlusion problem can be avoided. These factors suggest that the axis model would be less influenced by the uncertainty caused by the corresponding points than the horizontal model.

2.3 Image Brightness

An image pattern is often a brightness pattern in which the brightness values of images represent different properties of a scene. To quantitatively describe image brightness, some physical parameters should first be introduced.

2.3.1 Luminosity

Radiometry measures the energy of electromagnetic radiation. A basic quantity in radiometry is a radiation flux (or radiation power), whose unit is W. Light is a kind of electromagnetic radiation. **Visible light** has a spectrum ranging from 400 nm to

700 nm. Photometry measures the energy of light radiation. In photometry, radiation power is measured by luminous flux, whose unit is lm.

2.3.1.1 Point Source and Extended Source

When the scale of a light source is sufficiently small or it is so distant from the observer that the eye cannot identify its form, it is referred to as a **point source**. The luminous intensity I of a point source Q emitted along a direction r is defined as the luminous flux in this direction within a unit of solid angle (its unit is sr), as shown in Figure 2.17(a).

Take a solid angle element $d\Omega$ with respect to the r-axis and suppose the luminous flux in $d\Omega$ is $d\Phi$. The luminous intensity that emits from the point source along the r direction is

$$I = \frac{d\Phi}{d\Omega} \qquad (2.63)$$

The unit of luminous intensity is cd (1 cd = 1 lm/sr).

Real light sources always have some finite emitting surfaces. Therefore, they are also referred to as **extended sources**. On the surface of an extended source, each surface element dS has a limited luminous intensity dI along the r direction, as shown in Figure 2.17(b). The total luminous intensity of an extended source along the r direction is the sum of the luminous intensities of all the surface elements.

2.3.1.2 Brightness and Illumination

In Figure 2.17(b), suppose the angle between the r direction and the normal direction N of surface dS is θ. When an observer looks along the r direction, the projected surface area is $dS' = dS\cos\theta$. The **brightness** B of a surface element dS, looked at along the r direction, is defined as the luminous intensity of the unit-projected surface area along the r direction. In other words, brightness B is the luminous flux emitted from a unit of solid angle by a unit-projected surface

$$B \equiv \frac{dI}{dS'} \equiv \frac{dI}{dS\cos\theta} \equiv \frac{d\Phi}{d\Omega\,dS\cos\theta} \qquad (2.64)$$

The unit of brightness is cd/m^2.

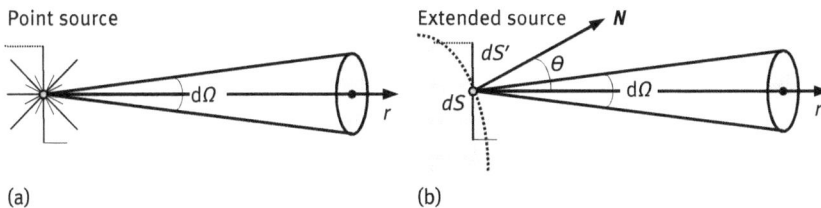

Figure 2.17: Point source and extended source of light.

The **illumination** of a surface illuminated by some light source is defined as the luminous flux on a unit surface area. It is also called irradiance. Suppose the luminous flux on a surface dS is $d\Phi$, then the illumination on this surface, E, is

$$E = \frac{d\Phi}{dS}$$ (2.65)

The unit of illumination is lx or lux ($1\,lx = 1\,lm/m^2$).

2.3.1.3 Factors Influencing the Density of an Image

There are a number of factors that determine the electrical strength of the signal forming the image of an object in a real scene (Joyce, 1985):

(1) Reflectivity: The reflectivity actually is what people are trying to measure. In the case of an object being viewed in transmitted light, both the thickness and the light-absorbency of the object will have an influence.

(2) Brightness of the light source and the efficiency of the optical train carrying the illumination to the specimen (e. g., condenser lenses, filters, and apertures in microscopic imaging systems).

(3) Absorption and reflection: In the image-forming part of the optical train, photons will be absorbed by the optical elements, and will also be reflected by various surfaces that they encounter. These reflections can end up in unwanted places, and the light from the particular portion of the object under consideration may be considerably enhanced by reflections arising from different parts of the object (glare).

(4) Conversion: The photo hits the light-sensitive surface of the acquiring devices (e. g., CCD) and its energy is converted to electrical energy in a linear or nonlinear manner.

(5) Amplification: The output from the acquiring device is then amplified, again in a nonlinear manner.

(6) Digitization: The signal is then digitized. The output of the digitizer may be passed through a look-up table converter in which the output has some predetermined function of the input.

The presence of these influencing factors means that the density of the images needs to be interpreted very carefully.

2.3.2 A Model of Image Brightness

An image can be taken as a 2-D bright function $f(x, y)$, where the brightness is a measurement of radiation energy. Therefore, $f(x, y)$ must be nonzero and finite

$$0 < f(x, y) < \infty$$ (2.66)

When capturing an image from a real scene, the brightness of the image is determined by two quantities: the amount of light incident to the viewing scene and the amount of light reflected by the object in the scene. The former is called the **illumination component** and is represented by a 2-D function $i(x, y)$. The latter is called the **reflection component** and is represented by a 2-D function $r(x, y)$. $i(x, y)$ is determined by the energy emitted by the light source and the distance between the source and the viewing scene (here, a point source is considered). $r(x, y)$ is calculated by the fraction of the reflection over incidence, which is determined by the surface property of the object. Some example values of typical surfaces are 0.01 for black velvet, 0.65 for stainless steel, 0.90 for a silver plate, and 0.93 for white snow. The value of $f(x, y)$ should be propositional to $i(x, y)$ and $r(x, y)$. This can be written as

$$f(x, y) = i(x, y)r(x, y) \tag{2.67}$$

According to the nature of $i(x, y)$ and $r(x, y)$, the following two conditions must be satisfied

$$0 < i(x, y) < \infty \tag{2.68}$$
$$0 < r(x, y) < 1 \tag{2.69}$$

Equation (2.68) means that the incident energy each time is greater than zero (only considering the case where radiation arrives at a surface) and it cannot go to infinite (for a physically realizable situation). Equation (2.69) means that the incident energy is always bound by 0 (total absorption) and 1 (total reflection).

The value of $f(x, y)$ is often called the gray-level value at (x, y), and can be denoted as g. Following eq. (2.67) to eq. (2.69), the gray-level values of $f(x, y)$ are also bound by two values: G_{min} and G_{max}. For images captured differently, both G_{min} and G_{max} vary. The restriction for G_{min} is that G_{min} must be positive if there is illumination. The restriction for G_{max} is that G_{max} must be finite. In a real application, the gray-level span $[G_{min}, G_{max}]$ is always converted to an integer range $[0, G]$. When an image is displayed, the pixel with $g = 0$ is shown as black and the pixel with $g = G$ is shown as white. All intermediate values are shown as shades from black to white.

2.4 Sampling and Quantization

Corresponding to the two parts in $f(x, y)$, the process of image acquisition consists of two parts:
(1) Geometry: Used to determine from where the image position (x, y) comes from in a 3-D scene.
(2) Radiometry (or photometry): Used to determine how bright a point (x, y) is in an image, and what the relationship of the brightness value with respect to the optical property of a 3-D point on the surface is.

When using an analogue image to obtain a digital image, two processes must be performed: sampling and quantization. The former establishes the size of an image (both the range of x and y), while the latter establishes the span values of f (the dynamic range of f).

2.4.1 Spatial Resolution and Amplitude Resolution

An image $f(x, y)$ must be digitized both in space and in amplitude to be processed by computers. The **sampling** is the process of the digitization of the spatial coordinates (x, y), and the **quantization** is the process of the digitization of the amplitude f.

Suppose that F, X, and Y are integer sets, $f \subset F$, $x \subset X$, $y \subset Y$. Sampling a continuous image can be accomplished by taking equally spaced samples in the form of a 2-D array. A spatially digitized image with a **spatial resolution** of $N \times N$ consists of N^2 pixels. Quantization of the above spatially digitized image can be accomplished by assigning equally spaced gray levels to each element in the image. The gray-level resolution (**amplitude resolution**, in general) is determined by the range of values for all pixels. An image with a gray-level resolution G has G distinct values.

For processing an image by computers, both N and G are taken as the power of 2, given by

$$N = 2^n \tag{2.70}$$

$$G = 2^k \tag{2.71}$$

Example 2.4 The spatial resolutions of several display formats
The spatial resolutions of several commonly used display formats are as follows:
(1) SIF (Standard Interface Format) in an NTSC system has the spatial resolution of 352×240. However, SIF in the PAL system has the spatial resolution of 352×288, which is also the spatial resolution of CIF (common intermediate format). QCIF (quarter common intermediate format) has a spatial resolution of 176×144.
(2) VGA: 640×480; CCIR/ITU-R 601: 720×480 (for NTSC) or 720×576 (for PAL); HDTV: 1440×1152 or 1920×1152.
(3) The screen of a (normal) TV has a length/high ratio of 4:3, while the screen of an HDTV is 16:9. To display an HDTV program on the screen of a (normal) TV, two formats can be used (as shown in Figure 2.18). One is the **double-frame format**, which keeps the original ratio. The other is the **whole-scan format**, which only intercepts a part of the original program along the horizontal direction. The former format retains the whole view with a reduced resolution; the latter format holds only a part but with the original resolution of this part. For example, suppose that a TV has the same height as an HDTV, it needs to receive the HDTV program with a spatial resolution of 1920×1080. If the double-frame format was used, the resolution would be 1440×810. If the whole scan format was used, the resolution would be 1440×1080. ▨

Figure 2.18: Displaying HDTV program on the screen of a normal TV.

The data space needed to store an image is also determined by spatial resolution and amplitude resolution. According to eq. (2.70) and eq. (2.71), the number of bits needed to store an image is

$$b = N^2 k \tag{2.72}$$

A digital image is an approximation of an analogue image. How many samples and gray levels are required for a good approximation? Theoretically, the larger the values of N and k, the better the approximation. Practically, the storage and the processing requirements will increase very quickly with the increase of N and k. Therefore, the values of N and k should be maintained on a reasonable scale.

Example 2.5 Storage and processing of image and video
A 512×512 image with 256 gray levels requires 2,097,152 bits of storage. A byte consists of 8 bits, so the above image needs 262,144 bytes of storage. A 1024×1024 color image needs 3.15 Mbytes to store. This requirement equals the requirement for a 750-page book. Video is used to indicate image sequence, in which each image is called a frame. Suppose that a color video has a frame size of 512×512, then the data volume for 1 second of video would be $512 \times 512 \times 8 \times 3 \times 25$ bits or 19.66 Mbytes.

To process a color video with a frame size of 1024×1024, it is needed to process $1024 \times 1024 \times 8 \times 3 \times 25 \approx 78.64$ Mbytes of data. Suppose that for each pixel ten floating-point operations (FLOPS) are required. One second of video needs nearly 1 billion FLOPS. Parallel processors can increase the process speed by simultaneously using many processors. The most optimistic estimation suggests that the processing time of a parallel process can be reduced to $(\ln J)/J$ with respect to that of a sequential process, where J is the number of parallel processors (Bow, 2002). According to this estimation, if 1 million processors are used to treat one second of video, each processor still needs to have the capability of nearly a hundred million FLOPS.

2.4.2 Image Quality Related to Sampling and Quantization

In the following, the relationship of **image quality** with respect to sampling and quantization is discussed. Spatial resolution and amplitude resolution are directly determined by sampling and quantization, respectively. Therefore, the subjective quality of images will degrade with respect to a decrease of the spatial resolution and the amplitude resolution of images. Three cases are studied here in which only the cases with sampling at equal intervals and uniform quantization are considered.

2.4.2.1 The Influence of Spatial Resolution

For a 512×512 image with 256 gray levels, if the number of gray levels are kept while reducing its spatial resolution (by pixel replication), a checkerboard effect with graininess will be produced. Such an effect is more visible around the region boundary in images and more important when the spatial resolution of images becomes lower.

Example 2.6 Effect of reducing spatial resolution

Figure 2.19 shows a set of images with different spatial resolutions. Figure 2.19(a) is a 512×512 image with 256 gray levels. Other figures are produced by keeping the gray levels while reducing the spatial resolution in both the horizontal and vertical directions to half of the previous image. The spatial resolution of Figure 2.19(b) is 256×256, Figure 2.19(c) is 128×128, Figure 2.19(d) is 64×64, Figure 2.19(e) is 32×32, and Figure 2.19(f) is 16×16.

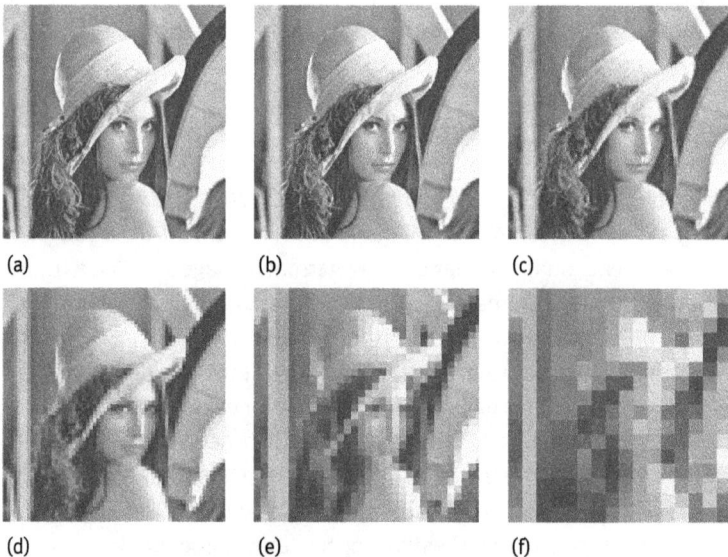

(a) (b) (c)

(d) (e) (f)

Figure 2.19: The effects of reducing the number of pixels in an image.

The effects induced by reducing the spatial resolution of images appear in different forms in these images. For example, the serration at the visor in Figure 2.19(b), the graininess of hairs in Figure 2.19(c), the blurriness of the whole image in Figure 2.19(d), the nearly unidentifiable face in Figure 2.19(e), and the hardly recognizable item in Figure 2.19(f).

2.4.2.2 The Influence of Amplitude Resolution

For a 512×512 image with 256 gray levels, keeping its spatial resolution while reducing its number of gray levels (by combining two adjacent levels to one), a degradation of the image quality will be produced. Such an effect is almost not visible when more than 64 gray levels are used. Further reducing of the number of gray levels will produce some ridge-like structures in images, especially around the areas with smooth gray levels. Such a structure becomes more sizeable when the number of gray levels is reduced. This effect is called false contouring and is perceived for an image displayed with 16 or less gray levels. Such an effect is generally visible in the smooth areas of an image.

Example 2.7 Effects of reducing amplitude resolution

Figure 2.19 shows a set of images with different amplitude resolutions. Figure 2.20(a) is the 512×512 image with 256 gray levels as in Figure 2.19(a). Other figures are produced by keeping the spatial resolution while reducing the number of gray levels. The number of gray levels is 64 in Figure 2.20(b), 16 in Figure 2.20(c), 8 in Figure 2.20(d), 4 in Figure 2.20(e), and 2 in Figure 2.20(f).

The effects of reducing the number of gray levels are hardly noticed in Figure 2.20(b), but they start to make an appearance in Figure 2.20(c). In Figure 2.20(d),

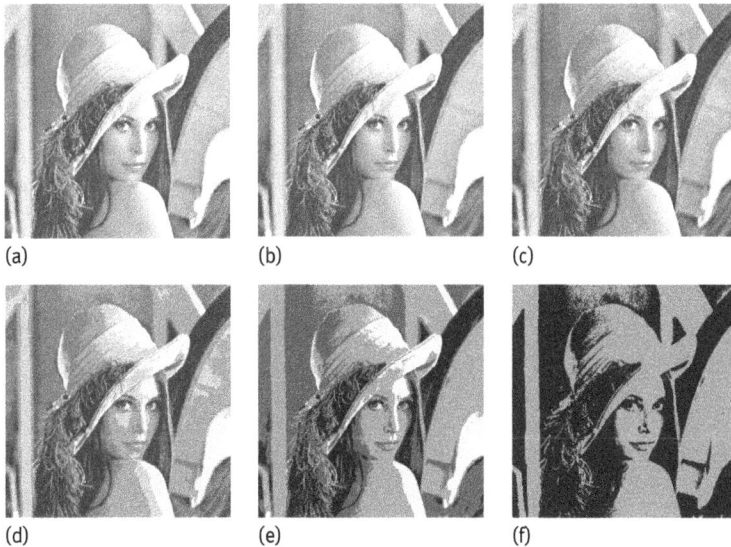

(a) (b) (c)

(d) (e) (f)

Figure 2.20: The effects of reducing the number of gray levels.

many false contours can be seen in the cap, shoulder, etc. Those effects are very noticeable in Figure 2.20(e), and Figure 2.20(f) looks like a woodcarving. ◎

2.4.2.3 The Influence of Spatial Resolution and Amplitude Resolution

The above two examples show the influences of spatial resolution and amplitude resolution, separately. Experiments consisting of subjective tests on images with varying **spatial resolutions** and **amplitude resolutions** have shown (Huang, 1965) the following:

(1) The quality of an image decreases with the reduction of spatial and amplitude resolutions. Only in a few cases where there is a fixed spatial resolution, will the reduction in amplitude resolution improve the quality of images.

(2) For images with many details, only a few numbers of gray levels are sufficient to represent them.

(3) For various images represented by the same number of gray levels, their subjective qualities can be quite different.

Example 2.8 Effect of reducing spatial and amplitude resolutions

Figure 2.21 shows a set of images with varying spatial and amplitude resolutions. Figure 2.21(a) is a 256 × 256 image with 128 gray levels. Figure 2.21(b) is a 181 × 181 image with 64 gray levels. Figure 2.21(c) is a 128×128 image with 32 gray levels. Figure 2.21(d) is a 90 × 90 image with 16 gray levels. Figure 2.21(e) is a 64 × 64 image with 8 gray levels, and Figure 2.21(f) is a 45 × 45 image with 4 gray levels.

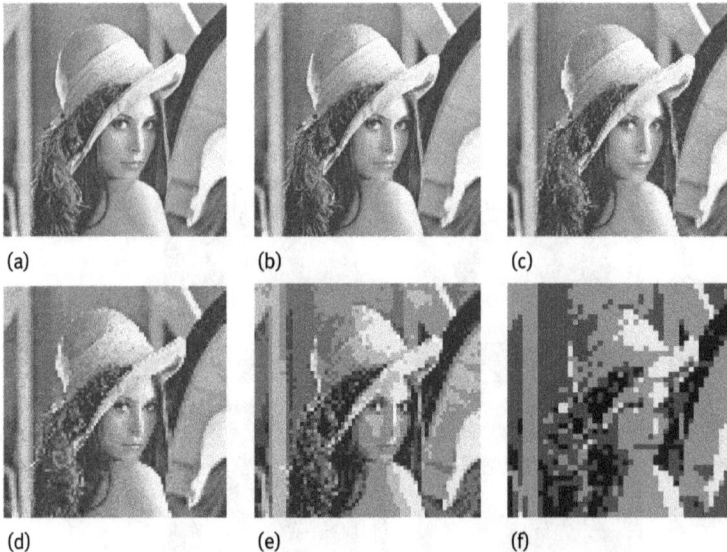

(a)   (b)   (c)

(d)   (e)   (f)

Figure 2.21: The effects of reducing both the number of pixels and gray levels.

Comparing Figure 2.21 to Figure 2.19 and Figure 2.20, it is seen that the degradation of the image quality is decreased more quickly if both the number of pixels and gray levels are reduced.

▣

2.4.3 Sampling Considerations

It is clear that sampling plays an important role in digital image acquisition. In the following, a more theoretical basis for this phenomenon will be discussed.

2.4.3.1 Sampling Theorem

Functions whose area under the curve is finite can be represented in terms of the sins and cosines of various frequencies (Gonzalez and Woods, 2002). The sine/cosine component with the highest frequency determines the highest "frequency content" of the functions. Suppose that this highest frequency is finite and that the function is of unlimited duration (these functions are called band-limited functions). Then, according to the **Shannon sampling theorem**, if the function is sampled at a rate (equal to or) greater than twice its highest frequency, it is possible to recover completely the original function from its samples.

Briefly stated, the **sampling theorem** tells us that if the highest frequency component in a signal $f(x)$ is given by w_0 (if $f(x)$ has a Fourier spectrum $F(w)$, $f(x)$ is band-limited to frequency $w0$ if $F(w) = 0$ for all $|w| > w_0$), the sampling frequency must be chosen such that $w_s > 2w_0$ (note that this is a strict inequality).

If the function is under-sampled, a phenomenon called aliasing corrupts the sampled image. The corruption is in the form of additional frequency components introduced into the sampled function. These are called aliased frequencies.

The sampling process can be modeled by

$$\hat{f}(x) = f(x) \sum_{n=-\infty}^{n=+\infty} \delta(x - nx_0) \tag{2.73}$$

which says that sampling is the multiplication of the signal $f(x)$ with an ideal impulse train with a spacing of x_0. The spacing is related to the sampling frequency by $w_s = 2\pi/x_0$.

Equation (2.73) can be rewritten as

$$\hat{f}(x) = \sum_{n=-\infty}^{n=+\infty} f(x)\delta(x - nx_0) = \sum_{n=-\infty}^{n=+\infty} f(nx_0)\delta(x - nx_0) \tag{2.74}$$

The set of samples can now be identified as $\{f_n\} = \{f(nx_0)|n = -\infty, \ldots, -1, 0, +1, \ldots, +\infty\}$. An infinite number of samples are necessary to represent a band-limited signal with the total fidelity required by the sampling theorem. If the signal were limited to some interval, say x_1 to x_2, then the sum in eq. (2.74) could be reduced to a finite sum with N samples where $N \approx (x_2 - x_1)/x_0$. The signal, however, could not be band-limited, and the sampling theorem could not be applied.

The above argument just shows that for any signal and its associated Fourier spectrum, the signal can be limited in extent (space-limited) or the spectrum can be limited in extent (band-limited) but not both.

The exact reconstruction of a signal from its associated samples requires an interpolation with the sinc(•) function (an ideal low-pass filter). With an impulse response $h(x)$, the reconstruction result is given by

$$f(x) = \sum_{n=-\infty}^{n=+\infty} f(nx_0)h(x - nx_0) = \sum_{n=-\infty}^{n=+\infty} f(nx_0)\frac{\sin(w_s(x - nx_0))}{w_s(x - nx_0)} \tag{2.75}$$

As it turns out, except for a special case discussed in the following paragraph, it is impossible to satisfy the sampling theorem in practice. People can only work with sampled data that are finite in duration. The process of converting a function of an unlimited duration into a function of a finite duration can be modeled simply by multiplying the unlimited function with a "gating function" that is valued as 1 for some interval and 0 elsewhere. Unfortunately, this function itself has frequency components that extend to infinity. Thus, the very act of limiting the duration of a band-limited function causes it to cease being band-limited (Young and Renswoude 1988).

There is one special case of significant importance in which a function of an infinite duration can be sampled over a finite interval without violating the sampling theorem. When a function is periodic, it may be sampled at a rate equal to or exceeding twice its highest frequency, and it is possible to recover the function from its samples provided that the sampling captures exactly an integer number of periods of the function.

2.4.3.2 Computer Memory Considerations

By taking examples from quantitative microscopy, it can show that the assumption that the collections of numbers in a computer memory faithfully represent the samples of a band-limited signal cannot be true.

(1) The input signal into a microscope is not band-limited; there is detail in the image down to any level.
(2) The image out of the microscope is almost band-limited; the MTF limits the spectrum of the output image to the bandwidth of the lens system.
(3) The band-limited signal is imaged on a surface, which is finite in extent, and thus only a finite portion of the signal is "observed" by the camera. The resulting signal from (out of) the camera is no longer band-limited.
(4) The image is sampled and the results are stored in a computer memory. Only a finite number of samples are used to represent the image. Thus, even if the camera had a photosensitive surface of an infinite spatial extent, the amount of data collected to represent the image would still be finite and the aliasing would be unavoidable.

It is now clear that applying the sampling theorem in a proper way is impossible due to either a finite camera aperture or the use of a finite amount data or both.

2.4.3.3 Measurements of Images

All of these arguments would be much less important for image processing than for image analysis. Now let us look at measurements of images. When measuring the property of an image, a collection of numbers in the computer memory, the digital image, is used as a representation of the true "analogue" image. A formula is applied to the digital image to generate an estimation of the analogue property.

The argument for using the sampling theory as a guide to choosing the proper sampling density is defective for two reasons.

(1)   The integrity of the information has already been compromised (put in danger) by limiting the sampled data to a finite number of samples (see above).

(2)   It is not possible to write a computer program that can extract an exact measure from the sampled data by executing a finite number of steps over a finite duration of time (see below).

To reconstruct the original analogue image, according to eq. (2.29), requires an infinite sum on n. Further, the interpolation terms (*i. e.*, the sinc($\cdot$) function) are themselves infinite in extent and converge relatively slowly to zero. A formula for extracting an exact measurement from an image requires all sampled data and thus includes an infinite number of terms. Further, if that formula also requires interpolated values of $f(x)$, an infinite number of sinc($\cdot$) functions must be evaluated to provide the correct value. Neither step can be performed in a finite number of computations.

It is clear that it is not possible to have an exact measurement and finite computation time. The central problem of digital image measurements then becomes the selection of the method that offers maximum accuracy with the minimum amount of computation.

Therefore, the proper choice of a sampling density cannot be based on as simple a criterion as the sampling theorem. High-accuracy measurements of analogue properties from the digital data can require considerable over-sampling (see Volume II of this book set).

2.5 Problems and Questions

2-1   What are the basic elements of human vision when considering the structure and functions?

2-2   (1) When an observer looks at a column 6 m high at a distance of 51 m, what is the size of the retinal image in millimeters?

(2) If changing the column to a 6 cm high column, at what distance does this column give the same retinal image size as in (1)?

2-3* What are the camera coordinates and image coordinates of a space point (1, 2, 3) after the perspective projection with a lens of focal length $\lambda = 0.5$?

2-4   Explain why inverse projection cannot uniquely map a point on the image plane to a 3-D point in the world coordinate system. Discuss what conditions should be satisfied to make this possible.

2-5   Suppose a camera is mounted in a position for surveillance as shown in Figure 2.8. The center of the camera is located at (0, 0, 1), the focal length of the camera is 0.5 m, the pan angle is 135°, and the tilt angle is also 135°. If a rectangular object is placed at a point (1 m, 1 m, 0.5 m) in the world coordinate system (the height of the object is along the Z axis), what are the minimum sizes (height and width) for the object to be detected?

2-6   When using a weak perspective projection, the problem of orientation ambiguity (target orientation) occurs when the target points in the image are not the same. List various ambiguities.

2-7   What is the difference between a sphere's perspective projection, the orthogonal projection, the weak perspective projection, and the parallel perspective projection? If the ball is replaced by an ellipsoid, what are the circumstances that need to be discussed separately?

2-8   One interesting phenomenon in everyday life is that street lamps located at different distances to the observer can have almost the same luminous intensity. Try to explain this phenomenon according to the photometry principle.

2-9*   If the output luminous flux of a street lamp is 2,000 lm, what is the brightness at the distances 50 m and 100 m, respectively?

2-10   Given a point source with a luminous flux equaling 3,435 lm, what is its luminous intensity? If a circle with a radius 2 cm is put at a location with the distance between the point source and the circle that equals 1 m, the light ray passing through the center of the circle has an angle with the normal of circle. What is the average illumination?

2-11   Suppose that the ratio of an image's width to its length is 4:3.

   (1) What is the spatial resolution of a mobile phone with a 300,000 pixel camera? If the camera has a resolution of 1,000,000 pixels, what is its spatial resolution?

   (2) What is the spatial resolution of a camera with 6,000,000 pixels? How many bits are needed to store a color image obtained with this camera?

2-12   Suppose that a TV has the same width as an HDTV and it is required to receive an HDTV program with a spatial resolution of 1920 × 1080. Compute the resolutions when using the double-frame format and the whole-scan format, respectively.

2.6 Further Reading

1. Spatial Relationship in Image Formation
   - Only basic perspective transformations are discussed in Section 2.1. More details on perspective transformations and other geometrical transformation,

such as the orthogonal transformation, can be found in Hartley and Zisserman (2004) and Gonzalez and Woods (2002).

2. **Stereo Imaging**
   - More discussions and more examples for various binocular imaging can be found in many books, for example, Shapiro and Stockman (2001), Hartley and Zisserman (2004), Szeliski (2010), and Davies (2012).

3. **Image Brightness**
   - The perception of brightness of an object and image depends not only on light sources and illumination but also on the psychology of observers. More introductions on visual perception can be found in Gibson (1950), Julesz (1960), Finkel and Sajda (1994), and Zakia (1997).
   - Radiometry and photometry are both well established in physics. Many textbooks for radiometry and photometry have been published, for example, Boyd (1983).

4. **Sampling and Quantization**
   - Sampling and quantization are discussed in many textbooks for image processing, for example, Gonzalez and Woods (1992) and Russ (2002).

3 Image Enhancement

Image enhancement is a group of image techniques used to improve the visual quality of images. These techniques may include pixel operations and mask operations, both of which can be linear or nonlinear. The techniques can be performed either in the image domain or in the frequency domain.

The sections of this chapter are arranged as follows:

Section 3.1 introduces some operations between images, including arithmetic operation and logic operation, and gives some examples of arithmetic operation in image enhancement.

Section 3.2 describes the principle of gray-level mapping, and some typical gray-level mapping functions and their image enhancement effects are also provided.

Section 3.3 focuses on the methods for histogram equalization and histogram specification. Histogram provides the statistical information of image, and re flects the gray-scale features and visual effects of image.

Section 3.4 describes the principle of frequency domain technique and the steps involved in frequency domain image enhancement. In addition, low-pass filtering and high-pass filtering are introduced, which are complementary in function.

Section 3.5 discusses some typical linear filtering methods. Their functions include smoothing and sharpening the image.

Section 3.6 introduces several typical nonlinear filtering methods. These methods may also be used in conjunction with linear filtering methods.

3.1 Image Operations

Image operations are often performed on the entire image; that is, every pixel in the image is processed in the same manner. In other words, the operand of such operations is an image, and the results of these operations can be obtained by processing all the pixels in the image.

3.1.1 Arithmetic and Logical Operations

Arithmetic and logical operations are basic image operations.

3.1.1.1 Arithmetic Operations
Arithmetic operations are often applied to gray-level images, pixel-by-pixel. The arithmetic operations between two pixels p and q include:
(1)   Addition: $p + q$.
(2)   Subtraction: $p - q$.

DOI 10.1515/9783110524116-003

(3) Multiplication: $p * q$ (which can be expressed by pq or $p \times q$).
(4) Division: $p \div q$.

The above operations take the values of the two pixels as the operands to compute the new value of the corresponding pixel in the output image. Sometimes the operation results may be out of the dynamic range of the gray levels. Therefore, certain mapping processes may be needed (see the next section).

3.1.1.2 Logic Operations
Logic operations are only applied to binary images, pixel by pixel. The principal logic operations include:
(1) COMPLEMENT: NOT q (also write as $\bar{q}$).
(2) AND: p AND q.
(3) OR: p OR q.
(4) XOR: p XOR q (which can also be written as $p \oplus q$. Different from the OR operation, when both p and q are 1, the XOR result is 0).

Example 3.1 Illustrations for the basic logic operations
Figure 3.1 shows some examples of logic operations, in which 1 is represented by the black color and 0 is represented by white.   ◎

By combining the above basic logic operations, various combined logic operations can be formed.

Example 3.2 Illustrations for the combined logic operations
Combining basic logic operations can produce other operations. Some results obtained using the images A and B in Figure 3.1 as the input are shown in Figure 3.2. The first row gives the results of (A) AND (NOT (B)), (NOT(A)) AND (B), (NOT(A)) AND (NOT(B)), and NOT$((A)$ AND$(B))$. The second row gives the results of (A) OR (NOT (B)),

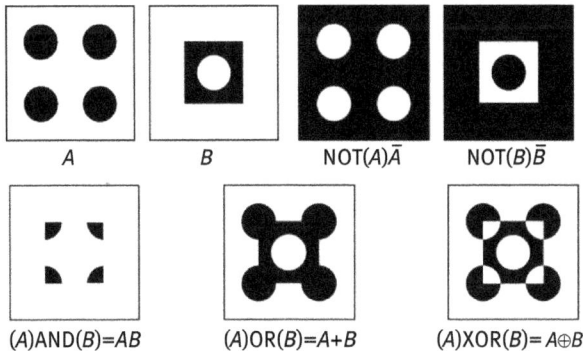

A ． B ． NOT$(A)\bar{A}$ ． NOT$(B)\bar{B}$

(A)AND$(B)=AB$ ． (A)OR$(B)=A+B$ ． (A)XOR$(B)=A\oplus B$

Figure 3.1: Illustrations of basic logic operations.

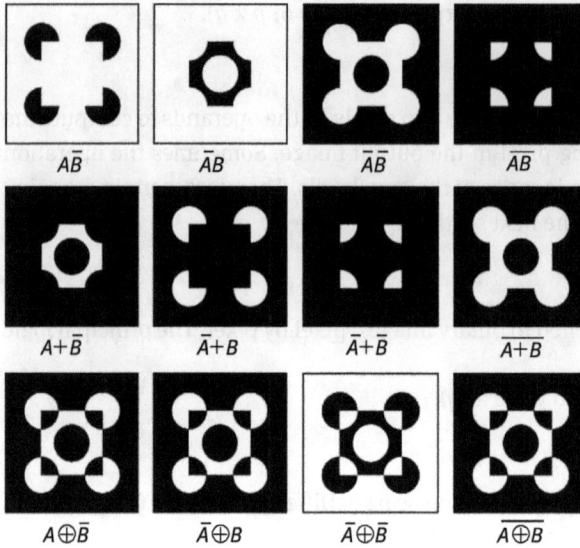

Figure 3.2: Illustrations of combined logic operations.

(NOT(A)) OR (B), (NOT(A)) OR (NOT(B)), and NOT((A) OR (B)). The third row gives the results of (A) XOR (NOT (B)), (NOT(A)) XOR (B), (NOT(A)) XOR (NOT(B)), and NOT((A) XOR(B)).

3.1.2 Applications of Image Operations

Image operations, such as arithmetic operations and the logic operations, are frequently used in image enhancement. A few examples are discussed in the following.

3.1.2.1 Applications of Image Addition
Image addition can be used to reduce or eliminate noise interfused during image acquisition. An acquired image $g(x, y)$ can be considered the superposition of the original image $f(x, y)$ and a noise image $e(x, y)$

$$g(x, y) = f(x, y) + e(x, y) \tag{3.1}$$

If the noise at different locations of the image are not correlated to each other and is of zero-mean, the noise can be eliminated by averaging a sequence of the acquired images. This process is represented by

$$\bar{g}(x, y) = \frac{1}{M} \sum_{i=1}^{M} g_i(x, y) \tag{3.2}$$

It is proven that the expectation of the new image is

$$E\{\bar{g}(x,y)\} = f(x,y) \tag{3.3}$$

The relation between the variance of the new image and the variance of the noised image is given by

$$\sigma_{\bar{g}(x,y)} = \sqrt{\frac{1}{M}} \times \sigma_{e(x,y)} \tag{3.4}$$

It can be seen that as M increases, the influence of the noise level at each pixel decreases.

Example 3.3 Eliminating random noise by averaging images
A set of images used to illustrate the noise removal by image averaging is shown in Figure 3.3. Figure 3.3(a) illustrates an image with 8-bit gray levels, where zero-mean Gaussian noise ($\sigma = 32$) is added. Figures 3.3(b), (c), and (d) are the results of using 4, 8, and 16 images for averaging. It is clear that with the increase in the number of images used for averaging, the influence of the noise is reduced progressively.   ◙

3.1.2.2 Applications of Image Subtraction
The subtraction of an image $h(x, y)$ from an image $f(x, y)$ reveals the difference between the two images

$$g(x,y) = f(x,y) - h(x,y) \tag{3.5}$$

Image subtraction is often used for background subtraction and motion detection. For a sequence of images, if the lighting condition can be considered fixed, the pixels with movements will have nonzero values in the difference image.

Example 3.4 Detecting object movement using image subtraction
In Figure 3.4, Figure 3.4(a)–(c) are three consecutive images of a sequence, Figure 3.4(d) is the difference image between Figures 3.4(a) and (b), Figure 3.4(e) is the

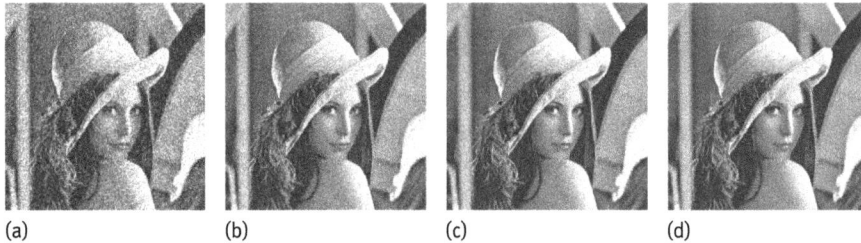

(a)   (b)   (c)   (d)

Figure 3.3: Elimination of random noise by image averaging.

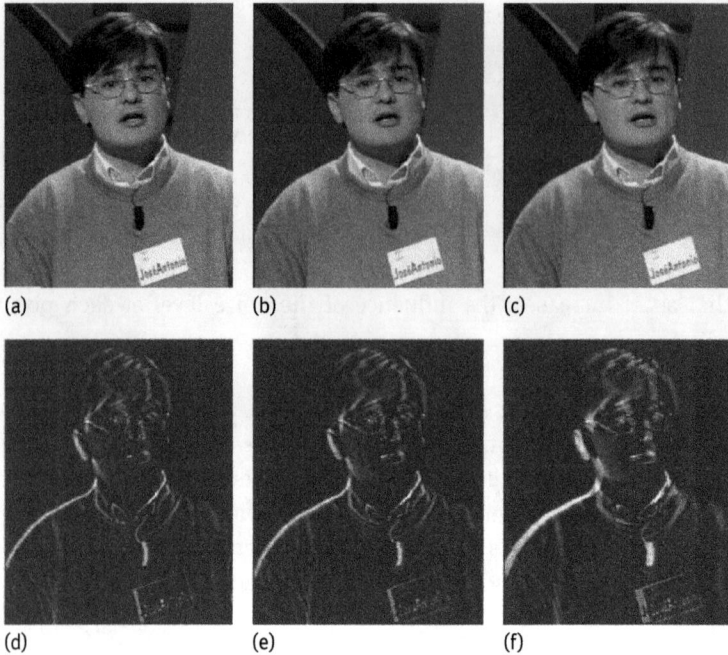

Figure 3.4: Detecting object movement using image subtraction.

difference image between Figure 3.4(b) and (c), and Figure 3.4(f) is the difference image between Figure 3.4(a) and (c).

3.1.2.3 Applications of Image Multiplication and Division

One important application of **image multiplication** (or **image division**) is to correct image brightness variations due to illumination or nonuniformity of the sensor. Figure 3.5 shows an example. Figure 3.5(a) gives a schematic image of the chessboard. Figure 3.5(b) shows the spatial variation of the scene brightness (nonuniformity). It is similar to the result of illumination when the light source is located above the upper left corner (in the absence of an object in the scene). Under this illumination, imaging Figure 3.5(a) will give the result shown in Figure 3.5(c), in which the lower right corner is the most unclear (with minimum contrast). If dividing Figure 3.5(c) by Figure 3.5(b), then the result of the correction of the illumination nonuniformity as shown in Figure 3.5(d) is obtained.

The explanation for the above-described treatment effect can be helped by means of the Figures 3.5(e)–(g). Figure 3.5(e) is a cross section of Figure 3.5(a), in which the gray level (vertical axis) as a function of the pixel coordinates (horizontal axis) has changed periodically. Figure 3.5(f) is a cross section of Figure 3.5(b), the gray level gradually reduces along the pixel coordinates. Figure 3.5(g) is the result of imaging Figure 3.5(e) under the illumination condition shown in Figure 3.5(f), the original

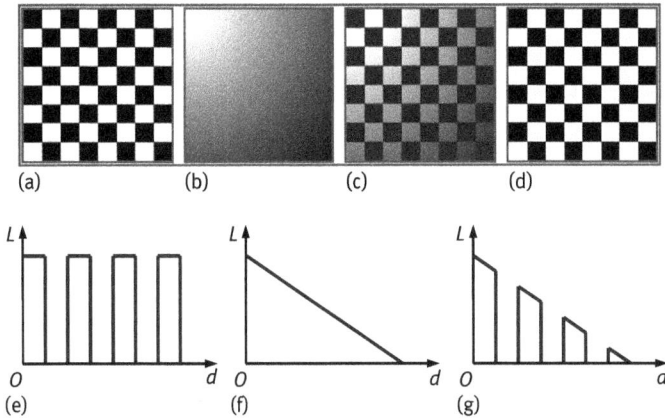

Figure 3.5: Uses image division to correct illumination nonuniformity.

flat platform is affected by the illumination and becomes a falling intermittent slope, the farther away from the origin, the smaller contrast. If dividing Figure 3.5(g) by Figure 3.5(f), it is possible to restore the same height of each platform of the Figure 3.5(e).

3.2 Direct Gray-Level Mapping

An image is composed of a number of pixels. The visual impression of an image is related to the gray-level value of each pixel. By modifying all or parts of the gray levels, the visual impression of an image can be changed.

3.2.1 Gray-Level Mapping

Direct gray-level mapping is a point-based operation (or single-point process), which means that according to some predefined mapping functions, every pixel in the original image will change its gray-level value to some other value. **Image enhancement** results will be obtained by assigning a new value to each pixel. The principle behind this enhancement method can be explained with the help of Figure 3.6. Suppose an image has four gray levels (R, Y, G, B), and the mapping function is shown in the middle of Figure 3.6. According to this mapping function, the original gray level G will be mapped to the gray level R and the original gray level B will be mapped to the gray level G. If the process is performed for all pixels, the left image will be transformed into the right image. By properly designing the form of the mapping functions, the required enhancement can be achieved.

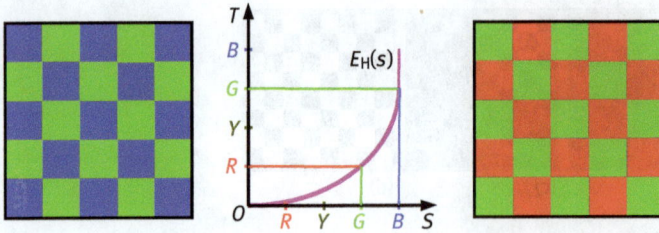

Figure 3.6: The principle behind direct gray-level mapping.

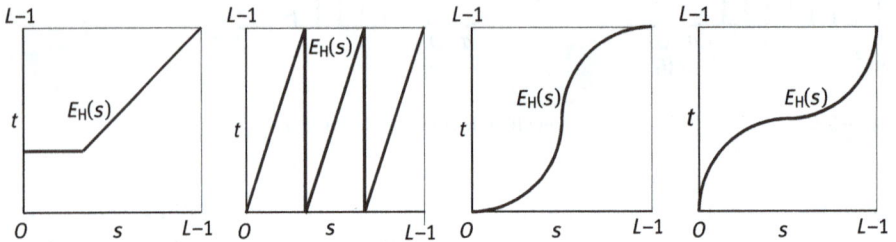

Figure 3.7: Illustration of various gray-level mapping functions.

The critical step of direct gray-level mapping is to design the **mapping function**, which is also called the transform function, according to enhancement requirements. Some examples of gray-level mapping functions are shown in Figure 3.7. Suppose the original gray-level value of a pixel is s and the mapped gray-level value of this pixel is t, then both the values are within the interval $[0, L - 1]$.

If the mapping function is a line from the origin to $(L-1, L-1)$, it has $s = t$. However, if the mapping function has the form of Figure 3.7(a), the pixels whose original gray-level values are smaller than the inflexion point of $E_H(s)$ will have the value of t. That is, all these pixels will have increased values. The mapping function in Figure 3.7(b) divides the original image into three parts according to the values of the pixels. In each part, the values of the pixels keep their order but are scaled to the range from 0 to $L-1$. Therefore, the contrast among the pixels of each part will increase. The bottom-left part of the mapping function in Figure 3.7(c) is similar to that in Figure 3.6, so the pixels with gray-level values less than $L/2$ in the original image will have even smaller values in the mapped image. However, the top-right part of the mapping function in Figure 3.7(c) is opposite to that in Figure 3.6, so the pixels with gray-level values bigger than $L/2$ in the original image will have higher values in the mapped image. The contrast of the whole image is thus increased. Finally, the mapping function in Figure 3.7(d) has some anti-symmetric properties with respect to the mapping function in Figure 3.7(c); this mapping function reduces the contrast of the image.

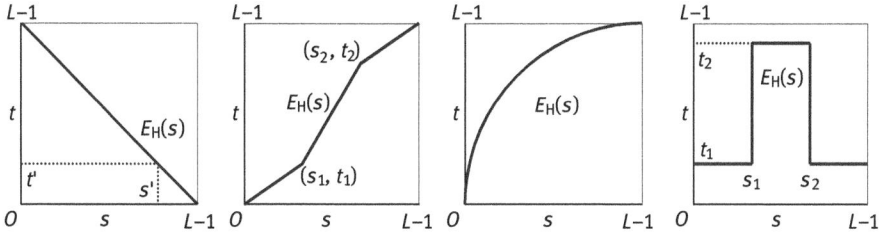

Figure 3.8: Typical examples of gray-level mapping functions.

3.2.2 Typical Examples of Mapping Functions

In real applications, various mapping functions can be designed to meet different requirements. Some typical examples of mapping functions are shown Figure 3.8.

3.2.2.1 Image Negatives

The negative of an image (**image negative**) is obtained by using the mapping function shown in Figure 3.8(a). The gray level s' will be mapped to the gray level t', so the original dark pixels will become brighter and the original bright pixels will become darker. The mapping here is done on a one-to-one basis. The process consists of reading the original gray-level value from a pixel, mapping it to a new gray level, and assigning this new value to the pixel.

3.2.2.2 Contrast Stretching

The idea behind **contrast stretching** is to increase the dynamic range of the gray levels in the whole image. One typical mapping function is shown in Figure 3.8(b), whose form is controlled by (s_1, t_1) and (s_2, t_2). Such a mapping reduces the dynamic ranges of $[0, s_1]$ and $[s_2, L-1]$, while it also increases the dynamic range of $[s_1, s_2]$. With different combinations of the values of s_1, s_2, t_1, and t_2, different effects can be produced. For example, if $t_1 = 0$ and $t_2 = L - 1$, then the mapping function has a slope larger than 1 and the original range of $[s_1, s_2]$ will occupy the whole range of $[0, L - 1]$ in the mapped image. If $s_1 = s_2$, $t_1 = 0$, and $t_2 = L - 1$, then only two levels are kept in the mapped image. In this case, the contrast is the biggest but most details are lost.

3.2.2.3 Dynamic Range Compression

In contrast to contrast stretching, in cases where the dynamic range of an image is out of the capability of the display device, some details of the image will be lost if this image is directly displayed. The solution for this problem is to compress the dynamic range of the original image (**dynamic range compression**) to fit the capability of the display device. One typical method is to use a logarithm mapping function (see Figure 3.8(c)):

$$t = C \log(1 + |s|) \tag{3.6}$$

where C is a scaling constant.

(a)   (b)   (c)   (d)

Figure 3.9: Results of direct gray-level mappings.

3.2.2.4 Gray-Level Slicing

The purpose of the **gray-level slicing** is similar to contrast stretching (*i.e.*, enhancing a certain gray level). One typical mapping function is shown in Figure 3.8(d), which highlights the range of $[s_1, s_2]$ and maps the remaining pixel values to some low gray levels.

Example 3.5 Results of direct gray-level mapping
Some illustrations of the results of the direct gray-level mapping are shown in Figure 3.9. Figures 3.9(a)–(d) correspond to the above four mappings (image negative, contrast stretching, dynamic range compression, and gray-level slicing). In Figure 3.9, the first line shows the original images, while the second line shows the results of the corresponding direct gray-level mappings. ◙

3.3 Histogram Transformation

The basis of image enhancement with histogram transformation is the probability theory. The basic techniques include histogram equalization and histogram specification.

3.3.1 Histogram Equalization

Histogram equalization is an automatic procedure for manipulating the histogram of images.

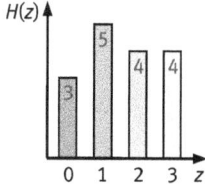

Figure 3.10: An image and its histogram.

3.3.1.1 Histogram and Cumulative Histogram

A **histogram** is a statistical representation of an image. For a gray-level image, its histogram provides the statistics about various gray levels in the image. Look at the image shown in Figure 3.10(a) and its gray-level histogram shown in Figure 3.10(b). In Figure 3.10(b), the horizontal axis represents various gray levels, while the vertical axis represents the number of pixels within each gray level.

Formally, the histogram of an image is a 1-D function

$$H(k) = n_k \quad k = 0, 1, \cdots, L - 1 \tag{3.7}$$

where n_k is the number of pixels with gray level k in $f(x, y)$. The histogram of an image gives an estimate of the probability of the occurrence of the gray level k in $f(x, y)$, and provides a global description of the appearance of the image.

Example 3.6 Images and corresponding histograms

Figure 3.11 shows a set of images and their histograms. Figure 3.11(a) is a high contrast image, whose histogram has a large dynamic range. Figure 3.11(b) is a low contrast image, whose histogram has only a narrow dynamic range. Figure 3.11(c) is a dark image, whose histogram has a large dynamic range but the gray levels are concentrated toward the dark side. Figure 3.11(d) is a bright image, whose histogram also has a large dynamic range but the gray levels are concentrated toward the bright side, in contrast to Figure 3.11(c). ◻

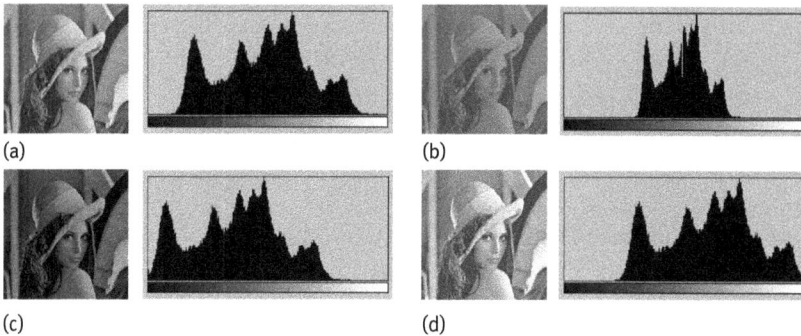

(a)

(b)

(c)

(d)

Figure 3.11: Different images and their histograms.

The **cumulative histogram** of an image is also a 1-D function

$$C(k) = \sum_{i=0}^{k} n_i \qquad k = 0, 1, \cdots, L - 1 \qquad (3.8)$$

In a cumulative histogram, the height of a bin k gives the total number of the pixels whose gray-level values are either smaller than or equal to k.

3.3.1.2 Histogram Equalization Function

Histogram equalization is often used to increase the contrast of an image with a small dynamic range. The idea behind histogram equalization is to transform the histogram into an evenly distributed form. In this way, the dynamic range of the image can be enlarged and the contrast of the image will be increased, too.

Write eq. (3.7) into a more general probability form as

$$p_s(s_k) = n_k/n \qquad \begin{array}{l} 0 \le s_k \le 1 \\ k = 0, 1, \cdots, L - 1 \end{array} \qquad (3.9)$$

where s_k is the value of the k-th gray level in $f(x, y)$, n is the total number of the pixels in $f(x, y)$, and $p(s_k)$ provides an estimation to the occurrence probability of s_k.

To enhance an image using the cumulative histogram, the following two conditions need to be satisfied:

(1) $E_H(s)$ is a single-valued and monotonically increased function for $0 \le s \le 1$. This condition preserves the order of the pixel values.

(2) Given $0 \le E_H(s) \le 1$, it has $0 \le s \le 1$. This condition guarantees the mapping results in the allowed dynamic range.

The inverse transformation $s = E_H^{-1}(t)$ also satisfies the above two conditions. The cumulative histogram can transform the distribution of s to an even distribution of t. In this case,

$$t_k = E_H(s_k) = \sum_{i=0}^{k} \frac{n_i}{n} = \sum_{i=0}^{k} p_s(s_i) \qquad \begin{array}{l} 0 \le s_k \le 1 \\ k = 0, 1, \cdots, L - 1 \end{array} \qquad (3.10)$$

3.3.1.3 Computing Histogram Equalization with Tables

In practice, histogram equalization can be performed step by step with the help of an equalization table.

Example 3.7 Computing histogram equalization with tables

Given an image of 64×64 with eight gray levels, whose histogram is shown in Figure 3.12(a). Figures 3.12(b) and (c) are the cumulative histogram and the equalized histogram, respectively. Note that the histogram equalization of a digital image is, in general, an approximation process, as it is impossible to map the pixels with the same value to different gray levels. This is clearly shown in Figure 3.12(d), when

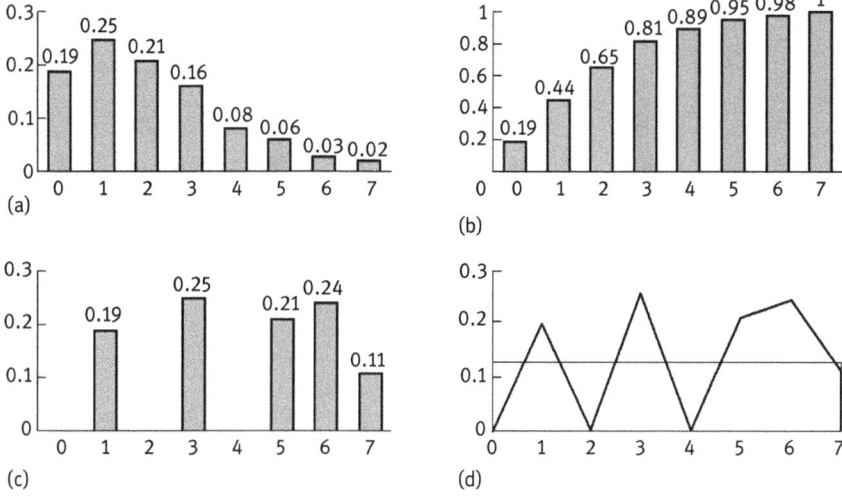

Figure 3.12: Histogram equalization.

Table 3.1: Computation of histogram equalization.

Label	Computation	Steps and results							
1	Original gray level k	0	1	2	3	4	5	6	7
2	Original histogram s_k	0.19	0.25	0.21	0.16	0.08	0.06	0.03	0.02
3	Compute t_k using eq. (3.10)	0.19	0.44	0.65	0.81	0.89	0.95	0.98	1.00
4	Take integer: $t_k = \text{int}[(L-1)t_k + 0.5]$	1	3	5	6	6	7	7	7
5	Determine mapping relation $(s_k \rightarrow t_k)$	$0 \rightarrow 1$	$1 \rightarrow 3$	$2 \rightarrow 5$	$3, 4 \rightarrow 6$		$5, 6, 7 \rightarrow 7$		
6	Equalized histogram		0.19		0.25		0.21	0.24	0.11

looking at the thick line (the real equalization result) and the horizontal line (the ideal equalization result). ◻

The steps and results for histogram equalization are summarized in Table 3.1.

Example 3.8 Results of histogram equalization
Example results for histogram equalization are shown in Figure 3.13.

Figures 3.13(a) and (b) give an 8-bit original image and its histogram, respectively. The dynamic range of the original image is narrow near the low end and the image looks somehow dark. The corresponding histogram is also narrow and all bins are concentrated on the low level side. Figures 3.13(c) and (d) present the results of the histogram equalization and the corresponding histogram, respectively. The new histogram completely occupies the dynamic range, and the new image has higher contrast and more details are revealed. Note that the histogram equalization increases the graininess (patchiness) as well as the contrast. ◻

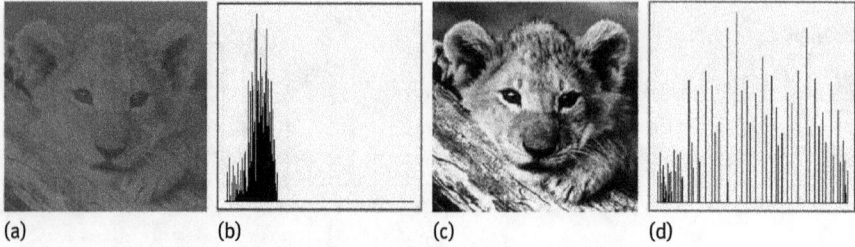

(a)   (b)   (c)   (d)

Figure 3.13: Results of histogram equalization.

3.3.2 Histogram Specification

Histogram equalization automatically increases the contrast of an image, but its effects are difficult to control. The image is always globally and evenly enhanced. In practical applications, sometimes it is needed to transform the histogram into a specific form to selectively enhance the contrast in a certain range or to make the distribution of the gray levels satisfy a particular requirement. In these cases, histogram specification can be used.

3.3.2.1 Three Steps of Histogram Specification

Histogram specification has three main steps (here, let M and N be the number of gray levels in the original image and the specified image, respectively, and $N \leq M$):

(1) Equalize the original histogram

$$t_k = E_{Hs}(s_i) = \sum_{i=0}^{k} p_s(s_i) \quad k = 0, 1, \cdots, M-1 \tag{3.11}$$

(2) Specify the desired histogram and equalize the specified histogram

$$v_l = E_{H_u}(u_j) = \sum_{j=0}^{l} p_u(u_j) \quad l = 0, 1, \cdots, N-1 \tag{3.12}$$

(3) Apply the inverse transform of step 1 and map all $p_s(s_i)$ to the corresponding $p_u(u_j)$.

The mapping function used in step 3 is critical when processing digital images. A frequently used mapping function is to search k and l, from smaller values to higher ones, which can minimize the following expression:

$$\left| \sum_{i=0}^{k} p_s(s_i) - \sum_{j=0}^{l} p_u(u_j) \right| \qquad \begin{matrix} k = 0, 1, \cdots, M-1 \\ l = 0, 1, \cdots, N-1 \end{matrix} \tag{3.13}$$

and then map $p_s(s_i)$ to $p_u(u_j)$. Since $p_s(s_i)$ is mapped one by one, this function is called the **single mapping law** (SML) (Gonzalez and Wintz, 1987). Using this law is quite simple, but it may produce more quantization errors.

A better method is the **group mapping law** (GML) (Zhang, 1992). Suppose it has an integer function $I(l), l = 0, 1, \ldots, N - 1$, that satisfies $0 \leq I(0) \leq \ldots \leq I(l) \leq \ldots \leq I(N - 1) \leq M - 1$. $I(l)$ can be determined to minimize the following expression

$$\left| \sum_{i=0}^{I(l)} p_s(s_i) - \sum_{j=0}^{l} p_j(u_j) \right| \qquad l = 0, 1, \cdots, N - 1 \qquad (3.14)$$

If $l = 0$, it maps all $p_s(s_i)$ whose i is in the range $[0, I(0)]$ to $p_u(u_0)$. If $l \geq 1$, it maps all $p_s(s_i)$ whose i is in the range $[I(l - 1) + 1, I(l)]$ to $p_u(u_j)$.

3.3.2.2 Computing Histogram Specification with Tables
Similar to the computation of histogram equalization with tables, the computation of histogram specification can also be performed with tables.

Example 3.9 Computing histogram specification with tables
The histogram in Figure 3.13(a) is used here. The computation steps and their results are shown in Table 3.2.

The corresponding histograms are shown in Figure 3.14. Figure 3.14(a) is the original histogram. Figure 3.14(b) is the (expected) specified histogram. Figure 3.14(c) shows the result obtained with SML. Figure 3.14(d) shows the result obtained with GML. The difference between Figures 3.14(b) and (c) is evident, while Figures 3.14(b)

Table 3.2: Computation of histogram specifications.

Label	Computation	Steps and results							
1	Original gray level k	0	1	2	3	4	5	6	7
2	Original histogram s_k	0.19	0.25	0.21	0.16	0.08	0.06	0.03	0.02
3	Computing the original cumulative histogram using eq. (3.10)	0.19	0.44	0.65	0.81	0.89	0.95	0.98	1.00
4	Specified histogram				0.2		0.6		0.2
5	Computing the specified cumulative histogram using eq. (3.10)				0.2	0.2	0.8	0.8	1.0
6S	SML mapping	3	3	5	5	5	7	7	7
7S	Determine mapping relation	0,1 → 3			2, 3, 4 → 5		5, 6, 7 → 7		
8S	Resulted histogram				0.44		0.45		0.11
6G	GML mapping	3	5	5	5	7	7	7	7
7G	Determine the mapping relation	0 → 3	1, 2, 3 → 5			4, 5, 6, 7 → 7			
8G	Resulted histogram				0.19		0.62		0.19

Note: Steps 6S to 8S are for SML mapping. Steps 6G to 8G are for GML mapping.

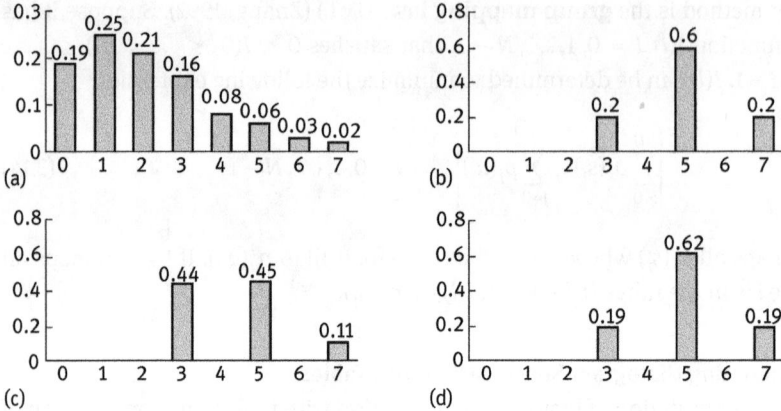

Figure 3.14: Histogram specification.

and (d) are quite similar. The superiority of GML over SML is obvious. Further examples can be found in Zhang (1992). ◎

Example 3.10 Results of histogram specification

The original image used here, as shown in Figure 3.15(a), is the same as in Example 3.9. As shown in Example 3.9, histogram equalization increases the contrast of the entire image but lacks detail in the dark regions. The histogram of the original image is shown in Figure 3.15(b) and it is used for histogram specification. The result image is shown in Figure 3.15(c), and the histogram is shown in Figure 3.15(d). Since the specified histogram has higher values in the bright regions, the processed image is brighter than that obtained by histogram equalization. In addition, the bins in the dark regions are more scattered, so the details in the dark regions become clear. ◎

3.3.2.3 Computing Histogram Specification with Drawings

In the computation of histogram specification, an intuitive and simple method is to use a drawing to determine the mapping relations. Here, a **drawing histogram**

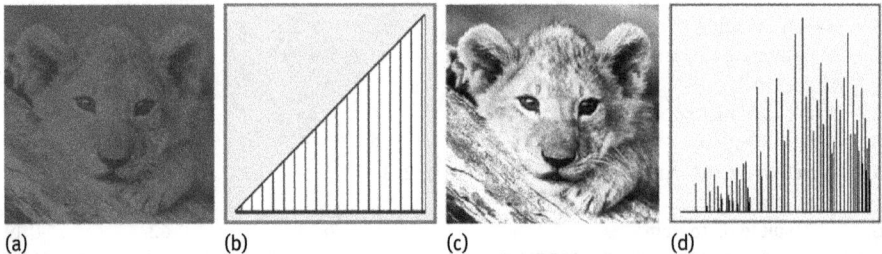

Figure 3.15: Results of histogram specification.

Original
cumulative

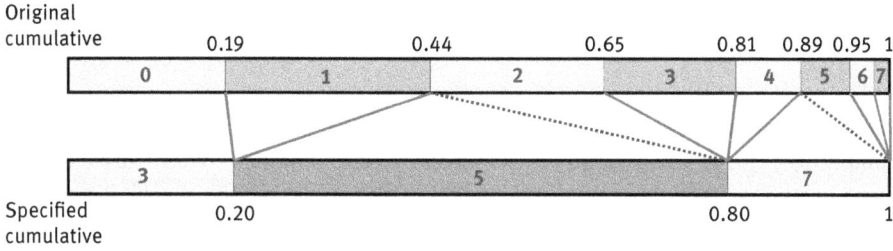

Figure 3.16: Illustration of SML.

depicts all bins of a histogram in a band, consecutively. Thus, the band represents a cumulative histogram and each segment of the band corresponds to a histogram bin.

In the SML process, the mapping is performed from the original cumulative histogram to the specified cumulative histogram. Each time, the shortest connection is taken. The data in Figure 3.16 are the same as in Figure 3.14. In Figure 3.16, 0.19 is mapped to 0.20, which is indicated by a solid line. Since the connection line between 0.44 and 0.20 (the solid line) is shorter than that between 0.44 and 0.80 (the dashed line), 0.44 is mapped to 0.20. Other connections are established similarly: 0.65 is mapped to 0.80, 0.81 is mapped to 0.80, 0.89 is mapped to 0.80, 0.95 is mapped to 1, 0.98 is mapped to 1, and 1 is mapped to 1. The ultimate result is the same as in Table 3.2.

In the GML process, the **mapping** is performed from the specified cumulative histogram to the original cumulative histogram. The shortest connection is also taken each time. The data in Figure 3.17 are the same as in Figure 3.16. In Figure 3.17, 0.20 is mapped to 0.19 as denoted by the solid line, but not to 0.44 as denoted by the dashed line. Similarly, 0.8 is mapped to 0.81 as denoted by the solid line, but not to 0.65 or 0.89 as denoted by the dashed line. After establishing these mapping relations, all bins in the original histogram can be mapped to the specified histogram. In Figure 3.17, the first bin of the original image is mapped to the first bin of the specified histogram; the second, third, and fourth bins are mapped to the second bin of the specified histogram; and the last four bins of the original histogram are mapped to the third bin of

Original
cumulative

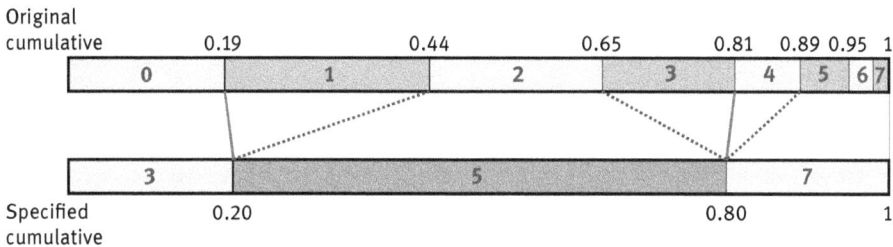

Figure 3.17: Illustration of GML.

the specified histogram. Thus, the obtained results, though described differently, are the same as in Table 3.2 and Figure 3.14(d).

3.3.2.4 A Comparison of SML and GML

Comparing Figures 3.16 and 3.17, the connection lines obtained by GML are closer to the vertical lines than those obtained by SML. This means that the specified cumulative histogram matches better to the original cumulative histogram. On the other hand, it is clear that SML is a biased mapping law since some gray levels are mapped toward the initial values of the computation, while GML is not biased.

A quantitative comparison can be made by counting the error introduced by mapping, which can be represented by the sum of the absolute difference between the original values and the mapped values. The smaller the sum, the better the mapping. Ideally, this sum should be zero. Consider again the data in Figures 3.16 and 3.17. The sum of the error for SML is $|0.44 - 0.20| + |(0.89 - 0.44) - (0.80 - 0.20)| + |(1 - 0.89) - (1 - 0.80)| = 0.48$, while the sum of the error for GML is $|0.20 - 0.19| + |(0.80 - 0.20) - (0.81 - 0.19)| + |(1 - 0.80) - (1 - 0.81)| = 0.04$. It is clear that the error introduced by GML is less than that introduced by SML.

Finally, look at the expected values of the error possibly introduced by these two laws. For the continuous case, both laws can provide accurate results. However, in the discrete case, their accuracy can be different. When mapping a $p_s(s_i)$ to a $p_u(u_j)$, the maximum error introduced by SML would be $p_u(u_j)/2$, while the maximum error introduced by GML would be $p_s(s_i)/2$. Since $N \leq M$, it is assured that $p_s(s_i)/2 \leq p_u(u_j)/2$. In other words, the expectation of the error for SML is greater than that for GML.

3.4 Frequency Filtering

In the frequency domain, the information from an image can be represented as a combination of various frequency components. Increasing or decreasing the magnitudes of certain frequency components will change the image in different ways.

The main steps for **frequency enhancement** are:
(1) Computing the Fourier transform of the original image $f(x, y)$, $T[f(x, y)]$.
(2) Enhancing $T[f(x, y)]$ in the frequency domain, $E_H[T[f(x, y)]]$.
(3) Computing the inverse Fourier transform of $E_H[T[f(x, y)]]$, that is, $T^{-1}\{E_H[T[f(x, y)]]\}$.

The whole process is given by

$$g(x, y) = T^{-1}\{E_H[T[f(x, y)]]\} \tag{3.15}$$

The enhancement in the frequency domain is made by multiplication of the Fourier transform with a transfer function (filter function). Transfer functions can remove or keep certain frequency components of the images to enhance images $H(u, v)$. The commonly used enhancement methods are low-pass filtering and high-pass filtering. Other methods, such as band-pass filtering and homomorphic filtering, can be obtained from low-pass filtering and high-pass filtering.

3.4.1 Low-Pass Filtering

Low-pass filtering will keep the low-frequency components and remove the high-frequency components. Since the noise in the gray levels of an image contributes most to the high-frequency components, low-pass filtering can reduce the effects of the noise.

3.4.1.1 Ideal Low-Pass Filters
A 2-D **ideal low-pass filter** has the transfer function

$$H(u, v) = \begin{cases} 1 & \text{if } D(u, v) \leq D_0 \\ 0 & \text{if } D(u, v) > D_0 \end{cases} \tag{3.16}$$

where D_0 is a nonnegative integer called the **cutoff frequency** and $D(u, v)$ is the distance from the point (u, v) to the center of frequency domain, $D(u, v) = (u^2 + v^2)^{1/2}$. A radial cross section of the ideal low-pass filter is shown in Figure 3.18. The cutoff frequency at D_0 always produces a ring effect (an effect making several contours around the structure in images, just like the waveforms produced around a bell when it is in the shake state) in a filtered image.

3.4.1.2 Butterworth Low-Pass Filters
A 2-D **Butterworth low-pass filter** of order n has the transfer function (the cutoff frequency is D_0)

$$H(u, v) = \frac{1}{1 + \left[\frac{D(u,v)}{D_0} \right]^{2n}} \tag{3.17}$$

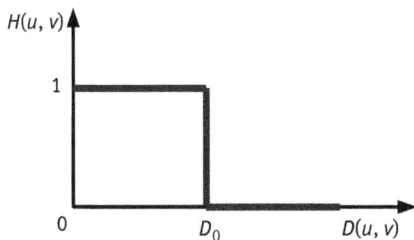

Figure 3.18: A radial cross section of the ideal low-pass filter.

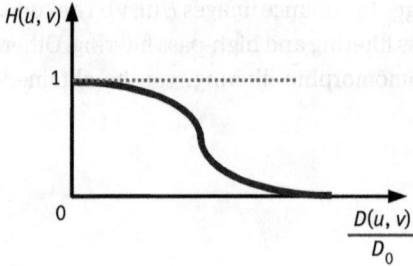

Figure 3.19 graph showing $H(u, v)$ on vertical axis with value 1 marked, curve descending, horizontal axis labeled $\frac{D(u, v)}{D_0}$, origin 0.

Figure 3.19: A radial cross section of the Butterworth low-pass filter of order 1.

A radial cross section of the Butterworth low-pass filter of order 1 is shown in Figure 3.19. It has no sharp discontinuity to establish a clear cutoff between the passed and the filtered frequencies. In this case, the cutoff frequency is often defined as the value of $H(u, v)$, which is less than a certain fraction of its maximum value. In eq. (3.16), when $D(u, v) = D_0$, $H(u, v) = 0.5$.

Example 3.11 Reducing the false contour with low-pass filtering
The false contour is caused by the use of an insufficient number of gray levels. This effect can be reduced by low-pass filtering. Figure 3.20(a) shows an image quantized to 12 gray levels. The false contour appears around the cap and the shoulder regions. Figures 3.20(b) and (c) show the filtered images, with the ideal low-pass filter and the Butterworth low-pass filter, respectively. Comparing these two images, it can be seen that although the false-contour effects are reduced in both images, there is an obvious ring effect in Figure 3.20(b), while Figure 3.20(c) is much better.   ▣

3.4.1.3 Other Low-Pass Filters
There are many types of low-pass filters. Two commonly used are trapezoid low-pass filters and exponential low-pass filters.

A 2-D **trapezoid low-pass filter** has the transfer function (the cutoff frequency is D_0)

Figure 3.20: Removing the false contour effect by low-pass filtering.

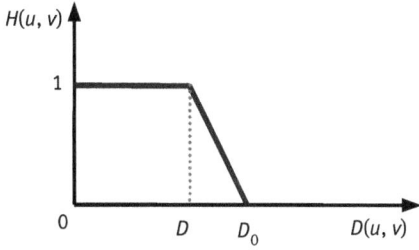

Figure 3.21: A radial cross section of the trapezoid low-pass filter.

$$H(u, v) = \begin{cases} 1 & \text{if} \quad D(u, v) \le D' \\ \dfrac{D(u, v) - D_0}{D' - D_0} & \text{if} \quad D' < D(u, v) < D_0 \\ 0 & \text{if} \quad D(u, v) > D_0 \end{cases} \tag{3.18}$$

where D' indicates the break point for the piecewise linear function. A radial cross section of the trapezoid low-pass filter of order 1 is shown in Figure 3.21. Compared to the transfer function of the ideal low-pass filter, it has a transition near the cutoff frequency and thus reduces the ring effect. On the other hand, compared to the transfer function of the Butterworth low-pass filter, its transition is not smooth enough, so its ring effect is stronger than that of the Butterworth low-pass filter.

A 2-D **exponential low-pass filter** of order n has the transfer function (the cutoff frequency is D_0)

$$H(u, v) = \exp\{-[D(u, v)/D_0]^n\} \tag{3.19}$$

A radial cross section of the exponential low-pass filter of order 1 is shown in Figure 3.22 (the order 2 filter is a Gaussian filter). It has a quite smooth transition between the low and high frequencies, so the ring effect is relatively insignificant (for the Gaussian filter, since the inverse Fourier transform of a Gaussian function is still a Gaussian function, no ring effect should occur). Compared to the transfer function of the Butterworth low-pass filter, the transfer function of the exponential low-pass filter decreases faster. Therefore, the exponential filter will remove more high-frequency components than the Butterworth filter does and causes more blurriness.

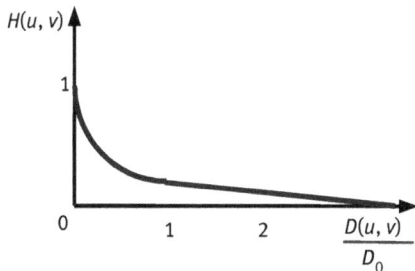

Figure 3.22: A radial cross section of the exponential low-pass filter of order 1.

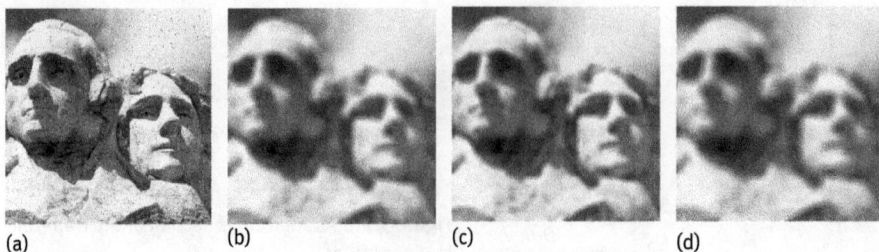

(a) (b) (c) (d)

Figure 3.23: A comparison *n* of the three low-pass filters.

Example 3.12 A comparison of the three low-pass filters
Figure 3.23(a) is a noisy image. Figures 3.23(b), (c), and (d) are the filtered images with the Butterworth filter, the trapezoid filter, and the exponential filter, respectively. All three filters can effectively remove noise and produce some ring effect. Among them, the blurriness is the strongest for the image filtered by the exponential filter as the filter removes most of the high-frequency components. The clearest image is obtained by the trapezoid filter as the number of the removed high-frequency components is the least using this filter.

3.4.2 High-Pass Filtering

High-pass filtering will keep the high-frequency components and remove the low-frequency components.

3.4.2.1 The Ideal High-Pass Filters
A 2-D **ideal high-pass filter** has the transfer function (see eq. (3.16))

$$H(u, v) = \begin{cases} 0 & \text{if } D(u, v) \le D_0 \\ 1 & \text{if } D(u, v) > D_0 \end{cases} \tag{3.20}$$

A radial cross section of the ideal high-pass filter is shown in Figure 3.24. It is complementary to that shown in Figure 3.18. The clear cutoff at D_0 also produces a ring effect in the filtered image.

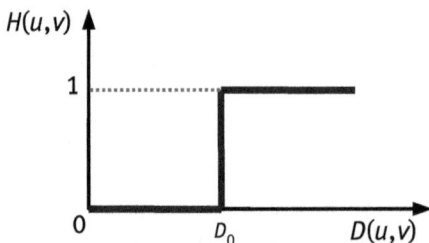

Figure 3.24: A radial cross section of the ideal high-pass filter.

$H(u, v)$

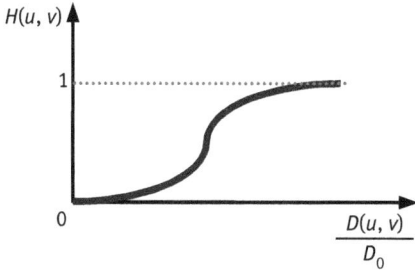

$\dfrac{D(u, v)}{D_0}$

Figure 3.25: A radial cross section of the Butterworth high-pass filter of order 1.

3.4.2.2 Butterworth High-Pass Filters

A 2-D **Butterworth high-pass filter** of order n has the transfer function (the cutoff frequency is D_0)

$$H(u, v) = \frac{1}{1 + \left[D_0 / D(u, v)\right]^{2n}} \tag{3.21}$$

A radial cross section of the Butterworth high-pass filter of order 1 is shown in Figure 3.25.

3.4.2.3 High-Frequency Emphasis Filters

High-pass filters often remove many low-frequency components (including the direct current component) and make the smooth regions darker or even completely black. To solve this problem, an offset is added to the transfer function to keep some low-frequency components. Such a filter is called a **high-frequency emphasis filter**.

Suppose that the Fourier transform of the original blurred image is $F(u, v)$ and the transfer function is $H(u, v)$, then the Fourier transform of the filtered image would be $G(u, v) = H(u, v)F(u, v)$. Adding a constant $c \in [0, 1]$ to $H(u, v)$, the transfer function of the high-frequency emphasis filter is derived as

$$H_e(u, v) = H(u, v) + c \tag{3.22}$$

The Fourier transform of the filtered image is

$$G_e(u, v) = G(u, v) + c \times F(u, v) \tag{3.23}$$

It includes high-frequency components and certain low-frequency components. Taking an inverse Fourier transform, it has

$$g_e(x, y) = g(x, y) + c \times f(x, y) \tag{3.24}$$

In this case, the enhanced image contains both the high-pass filtered results and a part of the original image. In other words, the enhanced image has more high-frequency components than the original image.

In practice, a constant k (k is greater than 1) is multiplied to the transfer function to further enhance the high-frequency components

$$H_e(u, v) = kH(u, v) + c \qquad (3.25)$$

Now eq. (3.24) becomes

$$G_e(u, v) = kG(u, v) + c \times F(u, v) \qquad (3.26)$$

Example 3.13 Result of high-frequency emphasis filtering
Figure 3.26(a) is a blurred image. Figure 3.26(b) shows the result of using the Butterworth high-pass filter. Since most of the low-frequency components are removed, the original smooth regions become very dark. Figure 3.26(c) shows the result of using the high-frequency emphasis filter ($c = 0.5$), where the edges are enhanced edges and more details of the image are preserved.

▣

3.4.2.4 High-Boost Filters

The high-pass filtering results can also be obtained by subtracting a low-pass filtered image from the original image. In addition, when subtracting a low-pass filtered image from the original image, which is multiplied by a constant A, a **high-boost filter** can be obtained. Suppose the Fourier transform of the original image is $F(u, v)$, the Fourier transform of the original image after the low-pass filtering is $F_L(u, v)$, and the Fourier transform of the original image after the high-pass filtering is $F_H(u, v)$, then the high-boost filtering result is

$$G_{HB}(u, v) = A \times F(u, v) - F_L(u, v) = (A - 1)F(u, v) + F_H(u, v) \qquad (3.27)$$

In eq. (3.27), $A = 1$ corresponds to the normal high-pass filter. When $A > 1$, a part of the original image is added to the high-pass filtered image. This creates a result closer to the original image. Comparing eq. (3.26) and eq. (3.27), it can be seen that when $k = 1$ and $c = (A - 1)$ the high-frequency emphasis filter becomes the high-boost filter.

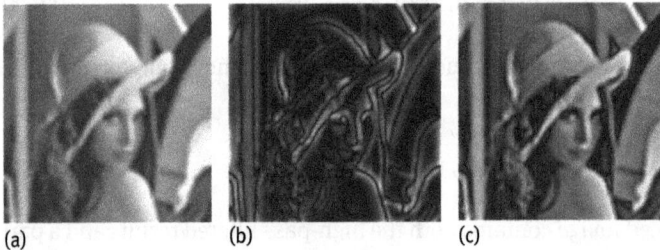

(a)   (b)   (c)

Figure 3.26: Comparison of high-pass filtering and high-frequency emphasis filtering.

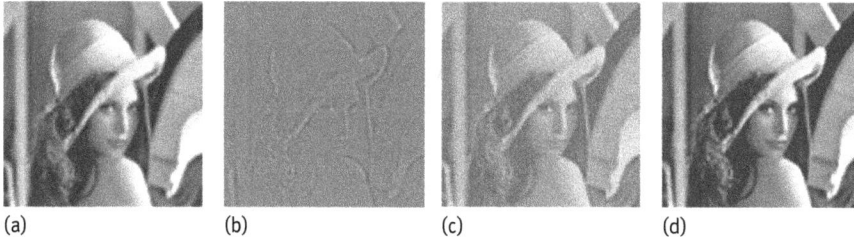

(a) (b) (c) (d)

Figure 3.27: Comparison of high-pass filtering and high-boost filtering.

Example 3.14 Comparing high-pass filtering with high-boost filtering

Figure 3.27(a) is a blurred image. Figure 3.27(b) gives the result obtained by high-pass filtering. As most of the low-frequency components are removed, the resulting image only has a small dynamic range. Figure 3.27(c) gives the result obtained by high-boost filtering (with $A = 2$), where a fraction of the low-frequency components has been recovered. Figure 3.27(d) gives the result obtained by further extending the dynamic range of Figure 3.27(c), which has a better contrast and clearer edges than the original image. ◉

3.5 Linear Spatial Filters

Image enhancement can be based on the property of the pixels and the relations between its neighbors. The techniques exploring these relations are often implemented with masks and are called spatial filtering.

3.5.1 Technique Classification and Principles

Spatial filters can be classified into two groups according to their functions.

3.5.1.1 Smoothing Filters
Smoothing filter can reduce or eliminate high-frequency components while keeping the low-frequency components of an image. Since the high-frequency components correspond to the part of the image with large and/or fast gray-level value changes, smooth filtering has the effect of reducing the local variation of the gray levels and makes an even appearance. In addition, smooth filtering can also conceal noise since noisy pixels have less correlation and correspond to the high-frequency components.

3.5.1.2 Sharpening Filters
Sharpening filter can reduce or eliminate low-frequency components while keeping the high-frequency components of an image. Since the low-frequency components correspond to the regions of small and slow gray-level changes, the sharpening filter

Table 3.3: Classification of spatial filtering techniques.

	Linear	Nonlinear
Smoothing	Linear smoothing	Nonlinear smoothing
Sharpening	Linear sharpening	Nonlinear sharpening

increases the local contrast and makes the edges sharper. In practice, the sharpening filter is often used to enhance blurred details.

On the other hand, **spatial filters** can also be classified into two groups: linear filters and nonlinear filters. From the statistical point of view, filters are estimators and they act on a group of observed data and estimate unobserved data. A linear filter provides a linear combination of the observed data, while a nonlinear filter provides a logical combination of the observed data (Dougherty and Astola, 1994). In linear methods, complicated computations are often decomposed into simple ones that can be performed quickly. Nonlinear methods provide better filtering results.

Combining the above two classification schemes, spatial filtering techniques can be classified into four groups: linear smoothing, nonlinear smoothing, linear sharpening, and nonlinear sharpening, as shown in Table 3.3. Only the smoothing techniques are presented below, the sharpening techniques will be discussed in Chapter 2 of Volume II in the section on edge detection.

Spatial filtering is performed with the help of masks (also called templates or windows). The idea behind the mask operation is that the new value of a pixel is the function of its old value and the values of the neighboring pixels. A mask can be viewed as an $n \times n$ (n is often a small odd constant) image.

Mask convolution includes the following steps:
(1) Shift the mask on an image and align the center of the mask with a pixel.
(2) Multiply the coefficients of the mask with the corresponding pixels in the image that are covered by the mask.
(3) Add all products together (this result is often divided by the number of coefficients needed to maintain the dynamic range).
(4) Assign the above result (the output of the mask) to the pixel at the center of the mask.

A more detailed description is shown in Figure 3.28. Figure 3.28(a) shows a portion of an image (the gray levels are represented by letters). Suppose a 3×3 mask will be used, whose coefficients are also represented by letters, as shown in Figure 3.28(b). If the center of the mask overlaps with the pixel marked by s_0, then the output of the mask R will be

$$R = k_0 s_0 + k_1 s_1 + \cdots + k_8 s_8 \tag{3.28}$$

The filtering of the pixel at (x, y) is completed by assigning R as its value, as shown in Figure 3.28(c).

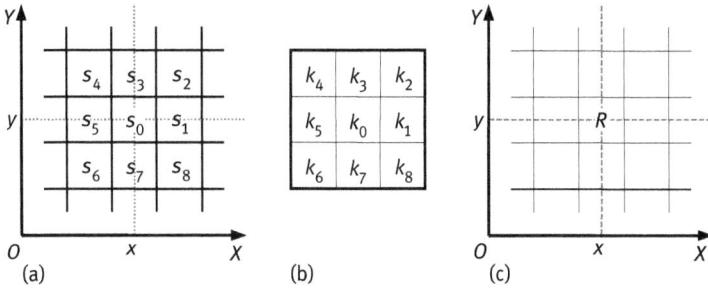

Figure 3.28: Illustration of partial filtering with a 3×3 mask.

3.5.2 Linear Smoothing Filters

Linear smoothing is one important type of spatial filtering technique.

3.5.2.1 Neighborhood Averaging

The simplest smoothing filter takes the mean value of the neighboring pixels as the output of the mask. In this case, all coefficients of the mask are 1. To keep the output image preserving the same dynamic range, the result obtained by eq. (3.28) should be divided by the number of coefficients. A simple averaging filter uses 1/9 to replace k (which is called the **box filter**).

Example 3.15 Illustration of neighborhood averaging
Figure 3.29(a) is an 8-bit image. Random noise is added to it as shown in Figure 3.29(b). Figures 3.29(c), (d), (e), (f), and (g) are the results of neighborhood averaging with

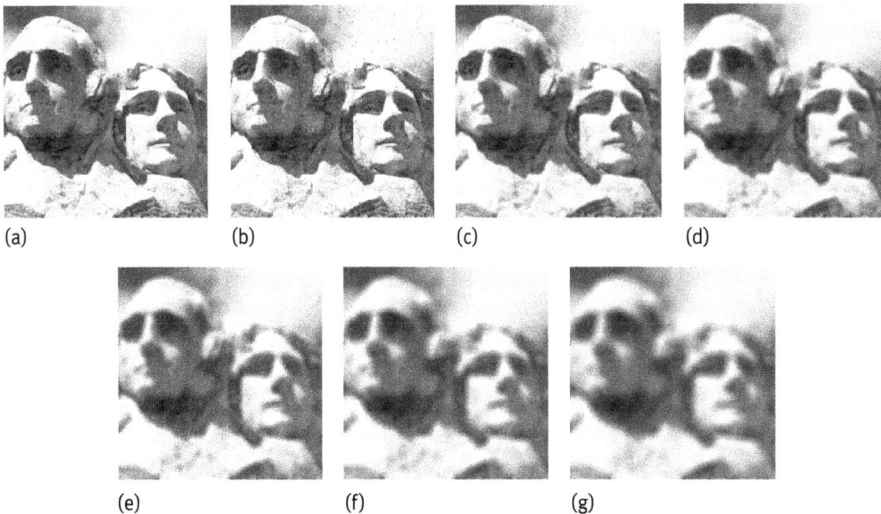

Figure 3.29: Results of neighborhood averaging.

masks of 3×3, 5×5, 7×7, 9×9, and 11×11, respectively. With the increase in the mask size, the performance of the noise removing process is increased. At the same time, the image also becomes more blurred. ▫

3.5.2.2 Weighted Average

In the mask, as shown in Figure 3.28(b), the coefficients can have different values. The values of k_0, k_1, ...,k_8 determine the function of the filter. To remove the noise, the values of the coefficients can be selected according to their distance to the center position. A common selection makes the values inversely proportional to the distance. For example, Figure 3.30 gives a **weighted smoothing mask**.

The coefficients of the mask can also be assumed to be Gaussian distributed, and thus the obtained filters are called the **Gaussian low-pass filters**, which are also weighted averaging filters. These filters are linear filters, which blur the images (smoothing the edges in an image) when reducing the noise. The computation of the coefficients can be made with the help of a Yanghui triangle, for the 1-D case. Table 3.4 lists the coefficients for a few small masks.

To obtain the discrete masks from the 1-D Gaussian function, the Gaussian function is sampled at the integer points $i = -n, \ldots, +n$

$$f(i) = \frac{1}{\sqrt{2\pi}} \exp(-i^2/2\sigma^2) \propto \exp(-i^2/2\sigma^2) \tag{3.29}$$

Let $n = 2\sigma + 1$, then the mask size is $w = 2n + 1$. For example, the maximum mask size is 7 for $\sigma = 1.0$ and is 11 for $\sigma = 2.0$.

1	2	1
2	4	2
1	2	1

Figure 3.30: A weighted smoothing mask.

Table 3.4: Coefficients of Gaussian masks.

$f(i)$	σ^2
1	0
1 1	1/4
1 2 1	1/2
1 3 3 1	3/4
1 4 6 4 1	1
1 5 10 10 5 1	5/4

The weighted average is known as the **normalized convolution** (Jähne, 1997). The normalized convolution is defined by

$$G = \frac{H \otimes (W \cdot F)}{H \otimes W} \tag{3.30}$$

where H is a convolution mask, F is the image to be processed, and W is the image with the weighting factors. A normalized convolution with the mask H essentially transforms the image F into a new image G.

The flexibility of the normalized convolution is given by the choice of the weighting image. The weighting image is not necessarily predefined. It can be determined according to the local structure of the image to be processed.

3.6 NonLinear Spatial Filters

Nonlinear spatial filters are often more complicated but also more powerful than linear spatial filters.

3.6.1 Median Filters

One of the commonly used nonlinear filters is the **median filter**.

3.6.1.1 Median Computation

The median filter, as its name implies, replaces the value of a pixel by the median of the gray-level values of the neighbors of that pixel (the original value of the pixel is also included in the computation of the median). This is given by

$$g_{median}(x, y) = \underset{(s,t)\in N(x,y)}{\text{median}} \{f(s, t)\} \tag{3.31}$$

where $N(x, y)$ is the neighborhood area (the filter mask) centered at (x, y).

The **median**, m, of a set of values satisfies such that half of the values in the set are less than or equal to m, while the rest are greater than or equal to m. In order to perform median filtering at a point in an image, the values of the pixel and its neighbors should be sorted. Their median is then determined and assigned to that pixel.

The principle function of median filters is to force points with distinct gray-level values to be similar to that of their neighbors. In fact, isolated clusters of pixels that are brighter or darker than that of their neighbors, and whose area is less than $n^2/2$ (one-half the filter area), are eliminated by an $n \times n$ median filter. Larger clusters are considerably less affected by median filters.

Median filtering can also be described by discrete distance. Suppose a discrete distance function d_D and a discrete distance value D are given, then a discrete mask $M(p)$

located at pixel p covers all pixels q such that $d_D(p, q) \leq D$. For example, if $d_D = d_8$ and $D = 1$, then $M(p) = \{q \text{ such that } d_8(p, q) \leq 1\}$ is the (3×3) mask centered at p.

Median filtering simply replaces the gray-level value at any pixel p by the median of the set of the gray-level values of the pixels contained in the mask $M(p)$. Given an ordered set S of N values, median(S) is the middle value of S. More precisely, if $S = \{q_1, \ldots, q_n\}$ and $q_i \geq q_j$ for $i > j$, median$(S) = q_{(N+1)/2}$ if N is odd, otherwise median$(S) = [q_{N/2} + q_{N/2+1}]/2$.

The values within the mask have to be sorted before the median value can be selected. Therefore, if $N_M = |M(p)|$, the median value can be determined in $O(N_M \log_2(N_M))$ operations. However, since the mask shifted from one pixel to another, the two masks corresponding to two neighboring pixels differ by $N_M^{1/2}$ values only. In this case, the median value can be determined in $O(N_M)$ operations for each pixel. In any case, this operation can be considered within a constant time (*i.e.*, it does not depend on the number of pixels N_{pixels} in the image), so that the overall complexity of median filtering is $O(N_{\text{pixels}})$.

3.6.1.2 Advantages of Median Filters

Median filters are popular for treating certain types of random noises. In fact, median filters are particularly effective in the presence of impulse noise. The median filter has the advantage of removing noise while keeping the original edges in the image. It can provide excellent noise-reduction capabilities, with considerably less blurriness than the linear smoothing filters of similar size. One example is given in Figure 3.31, where Figures 3.31(a) and (c) show the results of the filtering using box filters with 3×3 and 5×5 masks, respectively, and Figures 3.31(b) and (d) show the results of the filtering using median filters with 3×3 and 5×5 masks, respectively. The differences are evident.

3.6.1.3 2-D Median Masks

The shape of the mask has a considerable effect on the 2-D median filtering operation. The most common masks are depicted in Figure 3.32. Of these basic filter masks, the square mask is least sensitive to image details. It filters out thin lines and cuts the corners of the edges. It also often produces annoying strips (regions having a constant gray-level value, a drawback of median filters). The cross mask preserves thin

(a) (b) (c) (d)

Figure 3.31: Comparison of blurring effects of average filters and median filters.

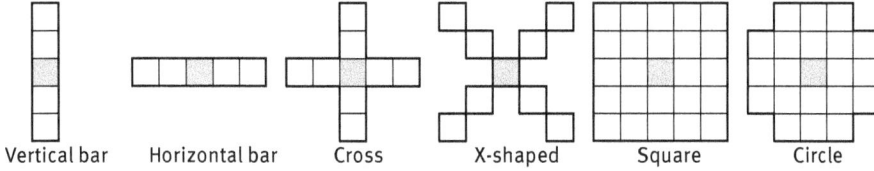

Figure 3.32: Some common masks used with the median filter.

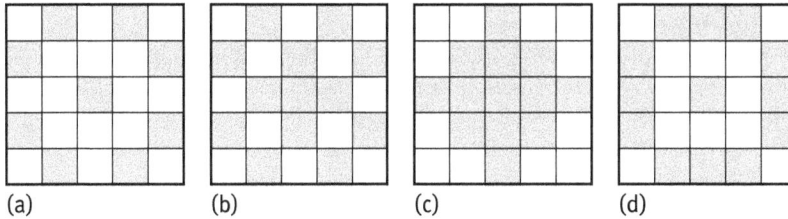

Figure 3.33: More masks used with the median filter.

vertical and horizontal lines but filters out diagonal lines, whereas the X-shaped mask preserves only diagonal lines. For humans, the use of the cross mask is more pleasing because horizontal and vertical lines are more important in human vision.

Other masks are depicted in Figure 3.33. The mask in Figure 3.33(a) uses 16-neighbors, while the mask in Figure 3.33(b) is obtained by combining the one in Figure 3.33(a) with four-neighbors. The mask in Figure 3.33(c) uses the pixels that have four-distances less or equal to two from the center pixel, while the mask in Figure 3.33(d) uses the pixels that have four-distances between 2 and 2.5 to the center pixel.

3.6.2 Order-Statistic Filters

Order-statistic filters are nonlinear spatial filters whose responses are based on the order (ranking) of the pixels contained in the image area encompassed by the filter. The value of the center pixel is replaced by the value determined by the ranking result. In these filters, instead of looking at the pixel values themselves, only the order of these pixel values is counted. The median filter is one of the best-known order-statistic filters (the median represents the 50th percentile of a set of ranked numbers). Other order-statistic filters are also used.

3.6.2.1 Max and Min Filters
The **max filter** uses the 100th percentile of a set of ranked numbers:

$$g_{max}(x, y) = \max_{(s,t) \in N(x,y)} \{f(s, t)\} \tag{3.32}$$

This filter is useful for finding the brightest points in an image. In addition, because the pepper noise has very low values, it can be reduced by this filter. This is a result of the maximum selection process in the neighborhood $N(x, y)$.

The 0th percentile filter is the **min filter**:

$$g_{min}(x, y) = \max_{(s,t) \in N(x,y)} \{f(s, t)\} \qquad (3.33)$$

This filter is useful for finding the darkest points in an image. In addition, it reduces the salt noise by the minimum operation.

Furthermore, the combination of the max and min filters can also be used to form a sharpening transform, which compares the maximum and minimum values with respect to the central pixel value in a small neighborhood (Ritter and Wilson, 2001),

$$S(x, y) = \begin{cases} g_{max}(x, y) & \text{if } g_{max}(x, y) - f(x, y) \le f(x, y) - g_{min}(x, y) \\ g_{min}(x, y) & \text{otherwise} \end{cases} \qquad (3.34)$$

This transform replaces the central pixel value by the extreme value in its neighbors, whichever is closest. This operation can be used as an (iterative) enhancement technique that sharpens fuzzy boundaries and brings fuzzy gray-level objects into focus (similar to the high-pass filter):

$$S^{n+1}(x, y) = S\{S^n(x, y)\} \qquad (3.35)$$

This transform also smooths isolated peaks or valleys (similar to the low-pass filter).

3.6.2.2 Midpoint Filter
The **midpoint filter** simply computes the midpoint between the maximum and minimum values in the area encompassed by the filter

$$g_{mid}(x, y) = \frac{1}{2} \left[\max_{(s,t) \in N(x,y)_-} \{f(s, t)\} + \min_{(s,t) \in N(x,y)} \{f(s, t)\} \right] \qquad (3.36)$$

Note that this filter combines the order statistic and averaging filters. This filter works best for randomly distributed noise, such as Gaussian or uniform noise.

3.6.2.3 Linear-Median Hybrid Filters
When the window size is large, the filters involve extensive computations and often become too slow for real-time applications. A way to solve this problem is to combine fast filters (in particular, linear filters) with order-statistic filters in such a way that the combined operation is reasonably close to the desired operation but is significantly faster (Dougherty and Astola, 1994).

In linear-median hybrid filtering, a cascade of the linear and median filters is used so that a small number of linear filters operate over large windows and the median of the outputs of the linear filters is used as the output of the **linear-median hybrid filter.**

Consider a 1-D signal $f(i)$. The linear-median hybrid filter with substructures $H_1, \ldots, H_M$ is defined by

$$g(i) = \text{MED}[H_1(f(i)), H_2(f(i)), \cdots, H_M(f(i))] \tag{3.37}$$

where $H_1, \ldots, H_M$ (M odd) are linear filters. The sub-filters H_i are chosen so that an acceptable compromise between noise reduction and achieving the root signal is set while keeping M small enough to allow simple implementation. As an example, look at the following structure:

$$g(i) = \text{MED}[H_L(f(i)), H_C(f(i)), H_R(f(i))] \tag{3.38}$$

The filter H_L, H_C, and H_R are low-pass filters following slower trends of the input signal. The filter H_C is designed to react quickly to signal-level changes, allowing the whole filter to move swiftly from one level to another. The subscripts L, C, and R represent the left, the center, and the right, which indicate the corresponding filter position with respect to the current output value, as shown schematically in Figure 3.34. The simplest structure consists of identical averaging filters H_L and H_R with $H_C[f(i)] = f(i)$. The whole filter is characterized by

$$g(i) = \text{MED}\left[\frac{1}{k} \sum_{j=1}^{k} f(i-k), f(i), \frac{1}{k} \sum_{j=1}^{k} f(i+k) \right] \tag{3.39}$$

The behavior of this filter is remarkably similar to the behavior of the standard median filter of length $2k + 1$ but computes quickly. In fact, using recursive-running sums, the complexity is constant and thus independent of the window size.

A commonly used number for sub-filters is five. For example, the filter

$$g(x, y) = \text{MED}\left\{ \frac{1}{2}[f(x, y-2) + f(x, y-1)], \frac{1}{2}[f(x, y+1) + f(x, y+2)], f(x, y) \right.$$
$$\left. \frac{1}{2}[f(x+2, y) + f(x+1, y)], \frac{1}{2}[f(x-1, y) + f(x-2, y)] \right\} \tag{3.40}$$

$f(i)$

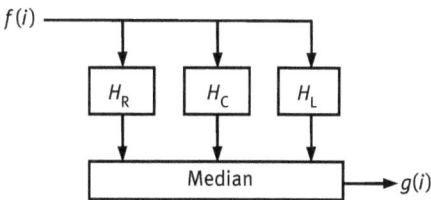

Figure 3.34: Computing scheme of the basic linear-median hybrid filter with sub-filters.

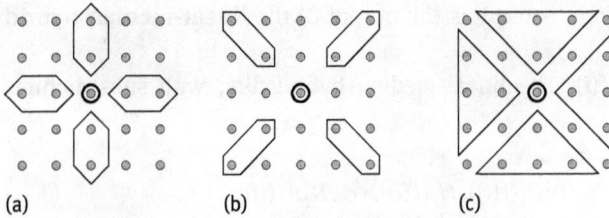

Figure 3.35: The masks for linear-median hybrid filters.

corresponds to the mask shown in Figure 3.36(a). Other typical mask shapes are shown in Figures 3.35(b) and (c).

3.7 Problems and Questions

3-1 Can an image be enhanced by only using geometric operations? Explain with examples.

3-2* How can you use logical operations only to extract the boundary of a rectangle in an image?

3-3 Select several gray-level images. Use the three lines given in Figure Problem 3-3 as the enhancement function to map these images. Compare their results and discuss.

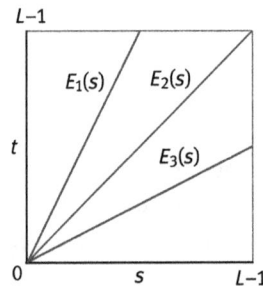

Figure Problem 3-3

3-4 In Figure Problem 3-4, $E_1(s)$ and $E_2(s)$ are two gray-level mapping curves:

(1) Explain the particularity, functionality, and suitable application areas of these two curves.

(2) Suppose that $L = 8$ and $E_1(s) = \text{int}[(7s)^{1/2} + 0.5]$. Map the image corresponding to the histogram shown in Figure 3.12(a) to a new image and draw the histogram of the new image.

(3) Suppose that $L = 8$ and $E_2(s) = \text{int}[s^2/7 + 0.5]$. Map the image corresponding to the histogram shown in Figure 3.12(a) to a new image and draw the histogram of the new image.

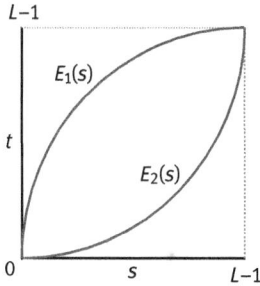

Figure Problem 3-4

3-5 Suppose that there is an $N \times N$ noiseless image, and the gray level in its left side is I, the gray level in its right side is J. Suppose that there is another $N \times N$ noiseless image, whose gray levels increase linearly from 0 in its left side to K in the right side ($K > J$). Multiply these two images to obtain a new image. What is the histogram of this new image?

3-6* Given an image with the histogram shown in Figure Problem 3-6(a), what is the result of the histogram specification with the specified histogram shown in Figure Problem 3-6(b)? After using both the SML method and the GML method, compare their errors.

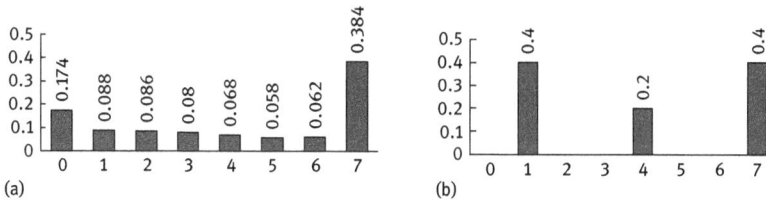

Figure Problem 3-6

3-7 A low-pass filter is formed by averaging the four-neighbors of a point (x, y), but excludes the point (x, y) itself. Obtain the filter transfer function, $H(u, v)$, which can perform the equivalent process in the frequency domain.

3-8 Show the differences between the ideal low-pass filter, Butterworth low-pass filter, trapezoid low-pass filter, and the exponential low-pass filter in image smoothing.

3-9 Why does repeatedly using a small mask obtain the effect of using a weighted large mask?

3-10 Considering an image with vertical strips of different widths, implement an algorithm:

(1) To smooth this image using the local smoothing in an eight-neighborhood.

(2) To smooth this image using the mask shown in Figure Problem 3.10 (dividing the result by 16).

1	2	1
2	4	2
1	2	1

Figure Problem 3.10

(3) Explain the different effects caused by using these two smoothing operations.

3-11 Add noise to the strip image with Gaussian noise, repeat Problem 3-10.

3-12 Adding M images and then taking the average can obtain the effect of reducing noise. Filtering an image with an $n \times n$ smoothing mask can also obtain the same effect. Compare the effects of these two methods.

3.8 Further Reading

1. **Image Operations**
   - Further details on noise sources and noise measurements can be found in Libbey (1994).
2. **Direct Gray-level Mapping**
   - Direct gray-level mapping is a basic and popular technique for image enhancement. More mapping laws can be found in Russ (2016).
3. **Histogram Transformation**
   - The original discussion on the group mapping law (GML) and its advantage over SML can be found in Zhang (1992).
   - Histogram equalization can be considered a special case of histogram specification (Zhang, 2004c).
4. **Frequency Filtering**
   - Frequency filtering for image enhancement has been discussed in many textbooks, such as (Gonzalez and Woods, 2008) and (Sonka *et al.*, 2008).
5. **Linear Spatial Filters**
   - Linear spatial filters are popularly used in practice. More methods for linear spatial filtering can be found, for example, in Pratt (2001) and Gonzalez and Woods (2008).
6. **NonLinear Spatial Filters**
   - Median filtering for bi-level images can be found in Marchand and Sharaiha (2000).
   - Median filter has been extended to switching median filter for performance improvement, see Duan and Zhang (2010).
   - More nonlinear spatial filters can be found in Mitra and Sicuranza (2001).
   - Linear spatial filters and nonlinear spatial filters can be combined to reduce various noise, one example is in Li and Zhang (2003).

4 Image Restoration

When acquiring and processing an image, many processes will cause degradation of the image and reduction of image quality. The image restoration process helps recover and enhance degraded images. Similar to image enhancement, several mathematic models for image degradation are often utilized for image restoration.

The sections of this chapter are arranged as follows:

Section 4.1 presents some reasons and examples of image degradation. The sources and characteristics of several typical noises and their probability density functions are analyzed and discussed.

Section 4.2 discusses a basic and general image degradation model and the principle for solving it. The method of diagonalization of circulant matrices is provided.

Section 4.3 introduces the basic principle of unconstrained restoration and focuses on the inverse filtering technique and its application to eliminate the blur caused by uniform linear motion.

Section 4.4 introduces the basic principle of constrained restoration and two typical methods: Wiener filter and constrained least square restoration.

Section 4.5 explains how to use the method of human–computer interaction to improve the flexibility and efficiency of image restoration.

Section 4.6 presents the two groups of techniques for image repairing. One group is image inpainting for restoring small-size area and another group is image completion for filling relative large-scale region.

4.1 Degradation and Noise

In order to discuss image restoration, some degradation factors, especially noise, should be introduced.

4.1.1 Image Degradation

There are several modalities used to capture images, and for each modality, there are many ways to construct images. Therefore, there are many reasons and sources for **image degradation**. For example, the imperfections of the optical system limit the sharpness of images. The sharpness is also limited by the diffraction of electromagnetic waves at the aperture stop of the lens. In addition to inherent reasons, blurring caused by defocusing is a common misadjustment that limits the sharpness in images. Blurring can also be caused by unexpected motions and vibrations of the camera system, which cause objects to move more than a pixel during the exposure time.

DOI 10.1515/9783110524116-004

Mechanical instabilities are not the only reason for image degradation. Other degradations result from imperfections in the sensor electronics or from electromagnetic interference. A well-known example is row jittering caused by a maladjusted or faulty phase-locked loop in the synchronization circuits of the frame buffer. This malfunction causes random fluctuations at the start of the rows, resulting in corresponding position errors. Bad transmission along video lines can cause echoes in images by reflections of the video signals. Electromagnetic interference from other sources may cause fixed movement in video images. Finally, with digital transmission of images over noisy wires, individual bits may flip and cause inhomogeneous, erroneous values at random positions.

The degradation due to the process of image acquisition is usually denoted as blurriness, which is generally band-limiting in the frequency domain. Blurriness is a deterministic process and, in most cases, has a sufficiently accurate mathematical model to describe it.

The degradation introduced by the recording process is usually denoted as **noise**, which is caused by measurement errors, counting errors, etc. On the other hand, noise is a statistical process, so sometimes the cause of the noise affecting a particular image is unknown. It can, at most, have a limited understanding of the statistical properties of the process.

Some degradation processes, however, can be modeled. Several examples are illustrated in Figure 4.1.

(1) Figure 4.1(a) illustrates nonlinear degradation, from which the original smooth and regular patterns become irregular. The development of the photography film can be modeled by this degradation.

(2) Figure 4.1(b) illustrates degradation caused by **blurring**. For many practically used optical systems, the degradation caused by the aperture diffraction belongs to this category.

(3) Figure 4.1(c) illustrates degradation caused by (fast) object motion. Along the motion direction, the object patterns are wiredrawing and overlapping. If the object moves more than one pixel during the capturing, the blur effect will be visible.

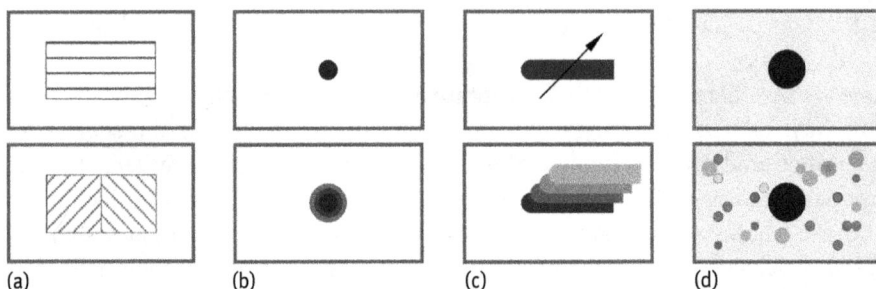

(a)   (b)   (c)   (d)

Figure 4.1: Four common examples of degradation.

(4)   Figure 4.1(d) illustrates degradation caused by adding noise to the image. This is a kind of random degradation, which will be detailed in the following. The original image has been overlapped by random spots that either darken or brighten the initial scene.

4.1.2 Noise and Representation

What is noise? **Noise** in general is considered to be the disturbing/annoying signals in the required signals. For example, "snow" on television or blurred printing degrades our ability to see and comprehend the contents. Noise is an aggravation.

Noise is an annoyance to us. Unfortunately, noise cannot be completely relegated to the arena of pure science or mathematics (Libbey, 1994). Since noise primarily affects humans, their reactions must be included in at least some of the definitions and measurements of noise. The spectrum and characteristics of noise determines how much it interferes with our aural concentration and reception. In TV, the black specks in the picture are much less distracting than the white specks. Several principles of psychology, including the Weber-Fechner law, help to explain and define the way that people react to different aural and visual disturbances.

Noise is one of the most important sources of image degradation. In image processing, the noise encountered is often produced during the acquisition and/or transmission processes. While in image analysis, the noise can also be produced by image preprocessing. In all cases, the noise causes the degradation of images.

4.1.2.1 Signal-to-Noise Ratio
In many applications, it is not important if the noise is random or regular. People are usually more concerned with the magnitude of the noise. One of the major indications of the quality of an image is its **signal-to-noise ratio** (SNR). The classical formula, derived from information theory, for the signal-to-noise ratio is given as

$$SNR = 10 \log_{10} \left(\frac{E_s}{E_n} \right) \tag{4.1}$$

This is actually stated in terms of a power ratio. Often, in specific scientific and technical disciplines, there may be some variations in this fundamental relationship. For example, the following defined SNR has been used in the process of generating images (Kitchen and Rosenfeld, 1981):

$$SNR = \left(\frac{C_{ob}}{\sigma} \right)^2 \tag{4.2}$$

where C_{ob} is the contrast between the object and background and σ is the standard deviation of the noise. More examples can be found in Chapter 8, where SNR is used as the objective fidelity criteria.

4.1.2.2 Probability Density Function of Noise

The **spatial noise descriptor** describes the statistical behavior of the gray-level values in the noise component of the image degradation model. These values come from several random variables, characterized by a **probability density function** (PDF). The PDF of some typical noises are described in the following.

Gaussian noise arises in an image due to factors such as electronic circuit noise and sensor noise which are caused by poor illumination and/or high temperature. Because of its mathematical tractability in both spatial and frequency domains, Gaussian (normal) noise models are used frequently in practice. In fact, this tractability is so convenient that the Gaussian models are often used in situations in which they are marginally applicable at best.

The PDF of a Gaussian random variable, z, is given by (see Figure 4.2)

$$p(z) = \frac{1}{\sqrt{2\pi\sigma}} \exp\left[-\frac{(z-\mu)^2}{2\sigma^2}\right] \tag{4.3}$$

where z represents the gray-level value, μ is the mean of z, and σ is its standard deviation (σ^2 is called variance of z).

Uniform Noise The uniform density is often used as the basis for numerous random number generators that are used in simulations. The PDF of the **uniform noise** is given by (see Figure 4.3)

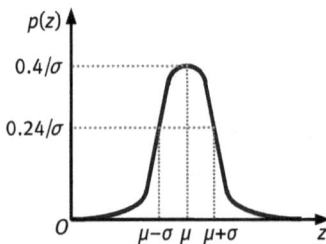

Figure 4.2: The PDF of a Gaussian noise.

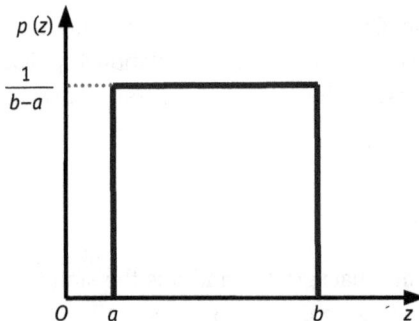

Figure 4.3: The PDF of a uniform noise.

$$p(z) = \begin{cases} 1/(b - a) & \text{if} & a \leq z \leq b \\ 0 & \text{otherwiser} \end{cases} \tag{4.4}$$

The mean and variance of this density are given by

$$\mu = (a + b)/2 \tag{4.5}$$
$$\sigma^2 = (b - a)^2/12 \tag{4.6}$$

Impulse Noise Impulse (salt and pepper) noise is found in situations where quick transient, such as faulty switching, takes place during image acquisition. Impulse noise also occurs in CMOS cameras with "dead" transistors that have no output or those that always output maximum values, and in interference microscopes for points on a surface with a locally high slope that returns no light (Russ, 2002). The PDF of (bipolar) **impulse noise** is given by (see Figure 4.4)

$$p(z) = \begin{cases} P_a & \text{for} & z = a \\ P_b & \text{for} & z = b \\ 0 & \text{otherwise} \end{cases} \tag{4.7}$$

If $b > a$, the gray-level value b will appear as a light dot in the image. Conversely, level a will appear like a dark dot. If either P_a or P_b is zero, the impulse noise is called unipolar. If neither probability is zero, and especially if they are approximately equal, impulse noise values will resemble salt-and-pepper granules randomly distributed over the image (its appearance as white and black dots is superimposed on an image). For this reason, bipolar noise is also called **salt-and-pepper noise** or **shot-and-spike noise**.

Noise impulse can be either negative or positive. Scaling usually is a part of the image digitizing process. Because impulse corruption is usually large compared with the strength of the image signal, impulse noise generally is digitized as the extreme (pure black or white) values in an image. Thus, it is usually assumed that a and b are "saturated" values, which take the minimum and maximum allowed values in the

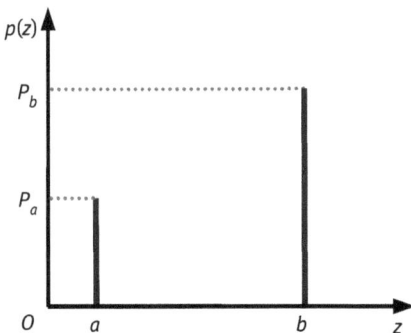

Figure 4.4: The PDF of an impulse noise.

digitized image. As a result, negative impulses appear as black (pepper) points in an image. For the same reason, positive impulses appear as white (salt) points. For an 8-bit image, this means that $a = 0$ (black) and $b = 255$ (white).

4.2 Degradation Model and Restoration Computation

Restoration computation depends on the model of degradation.

4.2.1 Degradation Model

A simple and typical **image degradation model** is shown in Figure 4.5. In this model, the degradation process is modeled as a system/operator H acting on the input image $f(x, y)$. It operates, together with an additive noise $n(x, y)$, to produce the degraded image $g(x, y)$. According to this model, the image restoration is the process of obtaining an approximation of $f(x, y)$, given $g(x, y)$ and the knowledge of the degradation operator H.

In Figure 4.5, the input and the output have the following relation

$$g(x, y) = H[f(x, y)] + n(x, y) \tag{4.8}$$

The degradation system may have the following four properties (suppose $n(x, y) = 0$):

4.2.1.1 Linear
Assuming that k_1 and k_2 are constants and $f_1(x, y)$ and $f_2(x, y)$ are two input images,

$$H[k_1 f_1(x, y) + k_2 f_2(x, y)] = k_1 H[f_1(x, y)] + k_2 H[f_2(x, y)] \tag{4.9}$$

4.2.1.2 Additivity
If $k_1 = k_2 = 1$, eq. (4.9) becomes

$$H[f_1(x, y) + f_2(x, y)] = H[f_1(x, y)] + H[f_2(x, y)] \tag{4.10}$$

Equation (4.10) says that the response of a sum of two inputs equals the sum of the two responses in a linear system.

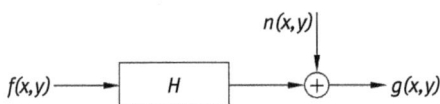

Figure 4.5: A simple image degradation model.

4.2.1.3 Homogeneity
If $f_2(x, y) = 0$, eq. (4.9) becomes

$$H\left[k_1 f_1(x, y)\right] = k_1 H\left[f_1(x, y)\right] \tag{4.11}$$

Equation (4.11) says that the response of the product of a constant and an input equals the product of the response of the input and the constant.

4.2.1.4 Position Invariant
If for any $f(x, y)$ and any a and b, there is

$$H\left[f(x - a, y - b)\right] = g(x - a, y - b) \tag{4.12}$$

Equation (4.12) says that the response of any point in the image depends only on the value of the input at that point, not the position of the point.

4.2.2 Computation of the Degradation Model

Consider first the 1-D case. Suppose that two functions $f(x)$ and $h(x)$ are sampled uniformly to form two arrays of dimensions A and B, respectively. For $f(x)$, the range of x is $0, 1, 2, \ldots, A - 1$. For $g(x)$, the range of x is $0, 1, 2, \ldots, B - 1$. $g(x)$ can be computed by convolution. To avoid the overlap between the individual periods, the period M can be chosen as $M \geq A + B - 1$. Denoting $f_e(x)$ and $h_e(x)$ as the extension functions (extending with zeros), their convolution is

$$g_e(x) = \sum_{m=0}^{M-1} f_e(m) h_e(x - m) \quad x = 0, 1, \cdots, M - 1 \tag{4.13}$$

Since the periods of both $f_e(x)$ and $h_e(x)$ are M, $g_e(x)$ also has this period. Equation (4.13) can also be written as

$$g = Hf = \begin{bmatrix} g_e(0) \\ g_e(1) \\ \vdots \\ g_e(M-1) \end{bmatrix} = \begin{bmatrix} h_e(0) & h_e(-1) & \cdots & h_e(-M+1) \\ h_e(1) & h_e(0) & \cdots & h_e(-M+2) \\ \vdots & \vdots & \ddots & \vdots \\ h_e(M-1) & h_e(M-2) & \cdots & h_e(0) \end{bmatrix} \begin{bmatrix} f_e(0) \\ f_e(1) \\ \vdots \\ f_e(M-1) \end{bmatrix} \tag{4.14}$$

From the periodicity, it is known that $h_e(x) = h_e(x + M)$. So, H in eq. (4.14) can be written as

$$H = \begin{bmatrix} h_e(0) & h_e(-1) & \cdots & h_e(1) \\ h_e(1) & h_e(0) & \cdots & h_e(2) \\ \vdots & \vdots & \ddots & \vdots \\ h_e(M-1) & h_e(M-2) & \cdots & h_e(0) \end{bmatrix} \tag{4.15}$$

H is a **circulant matrix**, in which the last element in each row is identical to the first element in the next row and the last element in the last row is identical to the first element in the first row.

Extending the above results to the 2-D case is direct. The extended $f_e(x)$ and $h_e(x)$ are

$$f_e(x, y) = \begin{cases} f(x, y) & 0 \le x \le A - 1 \quad \text{and} \quad 0 \le y \le B - 1 \\ 0 & A \le x \le M - 1 \quad \text{or} \quad B \le y \le N - 1 \end{cases} \tag{4.16}$$

$$h_e(x, y) = \begin{cases} h(x, y) & 0 \le x \le C - 1 \quad \text{and} \quad 0 \le y \le D - 1 \\ 0 & C \le x \le M - 1 \quad \text{or} \quad D \le y \le N - 1 \end{cases} \tag{4.17}$$

The 2-D form corresponding to eq. (4.13) is

$$g_e(x, y) = \sum_{m=0}^{M-1} \sum_{n=0}^{N-1} f_e(m, n) h_e(x - m, y - n) \quad \begin{matrix} x = 0, 1, \cdots, M - 1 \\ y = 0, 1, \cdots, N - 1 \end{matrix} \tag{4.18}$$

When considering the noise, eq. (4.18) becomes

$$g_e(x, y) = \sum_{m=0}^{M-1} \sum_{n=0}^{N-1} f_e(m, n) h_e(x - m, y - n) + n_e(x, y) \quad \begin{matrix} x = 0, 1, \cdots, M - 1 \\ y = 0, 1, \cdots, N - 1 \end{matrix} \tag{4.19}$$

Equation (4.19) can be expressed in matrices as

$$g = Hf + n = \begin{bmatrix} H_0 & H_{M-1} & \cdots & H_1 \\ H_1 & H_0 & \cdots & H_2 \\ \vdots & \vdots & \ddots & \vdots \\ H_{M-1} & H_{M-2} & \cdots & H_0 \end{bmatrix} \begin{bmatrix} f_e(0) \\ f_e(1) \\ \vdots \\ f_e(MN - 1) \end{bmatrix} + \begin{bmatrix} n_e(0) \\ n_e(1) \\ \vdots \\ n_e(MN - 1) \end{bmatrix} \tag{4.20}$$

where each H_i is constructed from the i-th row of the extension function $h_e(x, y)$,

$$H_i = \begin{bmatrix} h_e(i, 0) & h_e(i, N - 1) & \cdots & h_e(i, 1) \\ h_e(i, 1) & h_e(i, 0) & \cdots & h_e(i, 2) \\ \vdots & \vdots & \ddots & \vdots \\ h_e(i, N - 1) & h_e(i, N - 2) & \cdots & h_e(i, 0) \end{bmatrix} \tag{4.21}$$

where H_i is a circulant matrix. Since the blocks of **H** are subscripted in a circular manner, **H** is called a **block-circulant matrix**.

4.2.3 Diagonalization

Diagonalization is an effective way to solve eq. (4.14) and eq. (4.20).

4.2.3.1 Diagonalization of Circulant Matrices

For $k = 0, 1, \ldots, M - 1$, the eigenvectors and eigenvalues of the circulant matrix H are

$$w(k) = \left[1 \; \exp\left(j\frac{2\pi}{M}k\right) \cdots \exp\left(j\frac{2\pi}{M}(M-1)k\right) \right] \tag{4.22}$$

$$\lambda(k) = h_e(0) + h_e(M-1)\exp\left(j\frac{2\pi}{M}k\right) + \cdots + h_e(1)\exp\left(j\frac{2\pi}{M}(M-1)k\right) \tag{4.23}$$

Taking the M eigenvectors of H as columns, an $M \times M$ matrix W can be formed

$$W = \left[w(0) \quad w(1) \quad \cdots \quad w(M-1) \right]. \tag{4.24}$$

The orthogonality properties of w assure the existence of the inverse matrix of W. The existence of W^{-1}, in turn, assures the linear independency of the columns of W (the eigenvectors of H). Therefore, H can be written as

$$H = WDW^{-1} \tag{4.25}$$

Here, D is a diagonal matrix whose elements $D(k, k) = \lambda(k)$.

4.2.3.2 Diagonalization of Block-Circulant Matrices

Define a matrix W of $MN \times MN$ (containing $M \times M$ blocks of $N \times N$ in size). The im-th partition of W is

$$W(i, m) = \exp\left(j\frac{2\pi}{M}im\right) W_N \qquad i, m = 0, 1, \cdots, M-1 \tag{4.26}$$

where W_N is an $N \times N$ matrix, whose elements are

$$W_N(k, n) = \exp\left(j\frac{2\pi}{N}kn\right) \qquad k, n = 0, 1, \cdots, N-1 \tag{4.27}$$

Using the result for circulant matrices yields (note H is a block-circulant matrix)

$$H = WDW^{-1}. \tag{4.28}$$

Furthermore, the transpose of H, denoted H^T, can be represented with the help of the complex conjugate of $D(D^*)$

$$H^T = WD^*W^{-1} \tag{4.29}$$

4.2.3.3 Effect of Diagonalization

In the 1-D case (without noise), substituting eq. (4.28) into eq. (4.20) and performing left multiplication of the two sides by W^{-1} yields

$$W^{-1}g = DW^{-1}f \tag{4.30}$$

The products $W^{-1}f$ and $W^{-1}g$ are both M-D column vectors, in which the k-th item of $W^{-1}f$ is denoted $F(k)$, given by

$$F(k) = \frac{1}{M} \sum_{i=0}^{M-1} f_e(i) \exp\left(-j\frac{2\pi}{M}ki\right) \quad k = 0, 1, \cdots, M-1 \tag{4.31}$$

Similarly, the k-th item of $W^{-1}g$ is denoted $G(k)$, given by

$$G(k) = \frac{1}{M} \sum_{i=0}^{M-1} g_e(i) \exp\left(-j\frac{2\pi}{M}ki\right) \quad k = 0, 1, \cdots, M-1 \tag{4.32}$$

$F(k)$ and $G(k)$ are the Fourier transforms of the extended sequences $f_e(x)$ and $g_e(x)$, respectively.

The diagonal elements of D in eq. (4.30) are the eigenvalues of H. Following eq. (4.23),

$$D(k, k) = \lambda(k) = \sum_{i=0}^{M-1} h_e(i) \exp\left(-j\frac{2\pi}{M}ki\right) = MH(k) \quad k = 0, 1, \cdots, M-1 \tag{4.33}$$

where $H(k)$ is the Fourier transforms of the extended sequences $h_e(x)$.

Combining eq. (4.31) to eq. (4.33) yields

$$G(k) = M \times H(k)F(k) \quad k = 0, 1, \cdots, M-1 \tag{4.34}$$

The right side of eq. (4.34) is the convolution of $f_e(x)$ and $h_e(x)$ in the frequency domain. It can be calculated with the help of FFT.

Now, consider the 2-D cases (with noise). Taking eq. (4.28) into eq. (4.20), and multiplying both sides by W^{-1},

$$W^{-1}g = DW^{-1}f + W^{-1}n \tag{4.35}$$

where W^{-1} is an $MN \times MN$ matrix and D is an $MN \times MN$ diagonal matrix. The left side of eq. (4.35) is an $MN \times 1$ vector, whose elements can be written as $G(0, 0), G(0, 1), \ldots, G(0, N-1); G(1, 0), G(1, 1), \ldots, G(1, N-1); \ldots; G(M-1, 0), G(M-1, 1), \ldots, G(M-1, N-1)$. For $u = 0, 1, \ldots, M-1$ and $v = 0, 1, \ldots, N-1$, the following formulas exist

$$G(u, v) = \frac{1}{MN} \sum_{x=0}^{M-1} \sum_{y=0}^{N-1} g_e(x, y) \exp\left[-j2\pi\left(\frac{ux}{M} + \frac{vy}{N}\right)\right] \tag{4.36}$$

$$F(u, v) = \frac{1}{MN} \sum_{x=0}^{M-1} \sum_{y=0}^{N-1} f_e(x, y) \exp\left[-j2\pi\left(\frac{ux}{M} + \frac{vy}{N}\right)\right] \tag{4.37}$$

$$N(u, v) = \frac{1}{MN} \sum_{x=0}^{M-1} \sum_{y=0}^{N-1} n_e(x, y) \exp\left[-j2\pi\left(\frac{ux}{M} + \frac{vy}{N}\right)\right] \tag{4.38}$$

$$H(u, v) = \frac{1}{MN} \sum_{x=0}^{M-1} \sum_{y=0}^{N-1} h_e(x, y) \exp\left[-j2\pi\left(\frac{ux}{M} + \frac{vy}{N}\right)\right] \tag{4.39}$$

The MN diagonal elements of D can be represented by

$$D(k, i) = \begin{cases} MN \times H\left(\left\lfloor \frac{k}{N} \right\rfloor, k \bmod N\right) & \text{if } i = k \\ 0 & \text{if } i \neq k \end{cases} \tag{4.40}$$

where $\lfloor \cdot \rfloor$ is a flooring function (see Section 3.2).

Combining eq. (4.36) to eq. (4.40) and merging MN into $H(u, v)$ gives

$$G(u, v) = H(u, v)F(u, v) + N(u, v) \qquad \begin{array}{l} u = 0, 1, \cdots, M - 1 \\ v = 0, 1, \cdots, N - 1 \end{array} \tag{4.41}$$

To solve eq. (4.20), only a few discrete Fourier transforms of size $M \times N$ are needed.

4.3 Techniques for Unconstrained Restoration

Image restoration can be classified into unconstrained restoration and constrained restoration categories.

4.3.1 Unconstrained Restoration

From eq. (4.20), it can obtain

$$n = g - Hf \tag{4.42}$$

When there is no *a priori* knowledge of n, an estimation of f, $\hat{f}$ is to be found to assure that $H\hat{f}$ approximates g in a least squares sense. In other words, it makes the norm of n as small as possible

$$\|n\|^2 = n^T n = \left\|g - H\hat{f}\right\|^2 = (g - H\hat{f})^T(g - H\hat{f}) \tag{4.43}$$

According to eq. (4.43), the problem of image restoration can be considered a problem of minimizing the following function

$$L(\hat{f}) = \left\|g - H\hat{f}\right\|^2 \tag{4.44}$$

To solve eq. (4.44), differentiation of L with respect to $\hat{f}$ is needed, and the result should equal the zero vector. Assuming that H^{-1} exists and letting $M = N$, the **unconstrained restoration** can be represented as

$$\hat{f} = (H^T H)^{-1} H^T g = H^{-1} (H^T)^{-1} H^T g = H^{-1} g \tag{4.45}$$

4.3.1.1 Inverse Filtering

Taking eq. (4.28) into eq. (4.45) yields

$$\hat{f} = H^{-1} g = (WDW^{-1})^{-1} g = WD^{-1} W^{-1} g \tag{4.46}$$

Pre-multiplying both sides of eq. (4.46) by W^{-1} yields

$$W^{-1} \hat{f} = D^{-1} W^{-1} g \tag{4.47}$$

According to the discussion about diagonalization in Section 4.2, the elements comprising eq. (4.47) may be written as

$$\hat{F}(u, v) = \frac{G(u, v)}{H(u, v)} \qquad u, v = 0, 1, \cdots, M - 1 \tag{4.48}$$

Equation (4.48) provides a restoration approach often called **inverse filtering**. When considering $H(u, v)$ as a filter function that multiplies $F(u, v)$ to produce the transform of the degraded image $g\,(x, y)$, the division of $G(u, v)$ by $H(u, v)$ as shown in eq. (4.48) is just an inverse filtering operation. The final restored image can be obtained using the inverse Fourier transform

$$\hat{f}(x, y) = \mathcal{F}^{-1} \left[\hat{F}(u, v) \right] = \mathcal{F}^{-1} \left[\frac{G(u, v)}{H(u, v)} \right] \qquad x, y = 0, 1, \cdots, M - 1 \tag{4.49}$$

The division of $G(u, v)$ by $H(u, v)$ may encounter problems when $H(u, v)$ approaches 0 in the UV plane. On the other hand, noise can cause more serious problems. Substituting eq. (4.41) into eq. (4.48) yields

$$\hat{F}(u, v) = F(u, v) + \frac{N(u, v)}{H(u, v)} \qquad u, v = 0, 1, \cdots, M - 1 \tag{4.50}$$

Equation (4.50) indicates that if $H(u, v)$ is zero or very small in the UV plane, the term $N(u, v)/H(u, v)$ could dominate the restoration result and make the result quite different from the expected result. In practice, $H(u, v)$ often drops off rapidly as a function of the distance from the origin of the UV plane, while the noise usually drops at a much slower speed. In such situations, a reasonable restoration is possible to a small region around the origin point.

4.3.1.2 Restoration Transfer Function

For real applications, the inverse filter is not taken as $1/H(u, v)$, but as a function of the coordinates u and v. This function is often called the restoration transfer function and is denoted $M(u, v)$. The image degradation and restoration models are jointly represented in Figure 4.6.

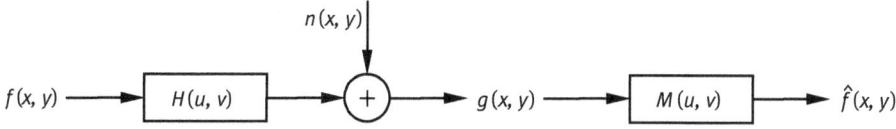

Figure 4.6: Image degradation and restoration models.

A commonly used $M(u, v)$ is given by

$$M(u, v) = \begin{cases} 1/H(u, v) & \text{if} \quad u^2 + v^2 \leq w_0^2 \\ 1 & \text{if} \quad u^2 + v^2 \leq w_0^2 \end{cases} \tag{4.51}$$

where w_0 is selected to eliminate the points at which $H(u, v)$ equals zero. However, the restoration results obtained by this method often have a visible "ring effect." An improved method is to take $M(u, v)$ as

$$M(u, v) = \begin{cases} k & \text{if} \quad H(u, v) \leq d \\ 1/H(u, v) & \text{otherwise} \end{cases} \tag{4.52}$$

where k and d are both constants that are less than 1.

Example 4.1 Blurring a point image to obtain the transfer function and restore the image

The transfer function $H(u, v)$ of a degraded image can be approximated using the Fourier transform of the degraded image. One full image can be considered a combined set of point images. If a point image is taken as the approximation of a unit impulse function ($\mathscr{F}[\delta(x, y)] = 1$), then $G(u, v) = H(u, v)F(u, v) \approx H(u, v)$.

Figure 4.7 illustrates the results of the restoration. Figure 4.7(a) is a simulated degradation image, which is obtained by applying a low-pass filter to an ideal image. The Fourier transform of the used low-pass filter is shown in Figure 4.7(b). The restoration results obtained using eq. (4.51) and eq. (4.52) are shown in Figures 4.7(c) and (d), respectively. Comparing these two results, the "ring effect" in Figure 4.7(d) is less visible. ▣

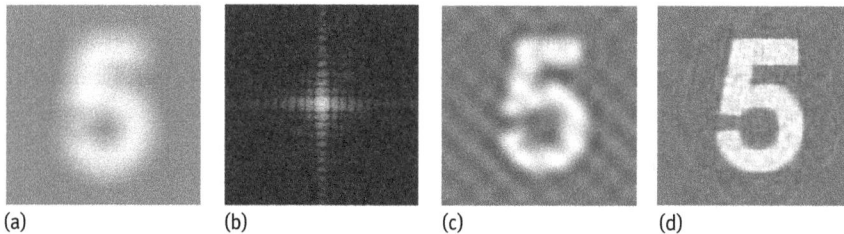

(a)　　　　　(b)　　　　　(c)　　　　　(d)

Figure 4.7: Examples of restoration.

4.3.2 Removal of Blur Caused by Uniform Linear Motion

In some applications, $H(u, v)$ can be obtained analytically. One example is the application used to restore blurred images, which is caused by uniform linear motion.

First, consider the continuous case. Suppose that an image $f(x, y)$ is captured for a uniform moving object on the plane and the moving components of the object in X and Y directions are denoted $x_0(t)$ and $y_0(t)$, respectively. The duration for image capturing is T. The real blurred image $g(x, y)$, which is caused by the motion is

$$g(x, y) = \int_0^T f[x - x_0(t), y - y_0(t)]dt \tag{4.53}$$

The Fourier transform of $g(x, y)$ is

$$G(u, v) = \int_{-\infty}^{\infty} \int_{-\infty}^{\infty} g(x, y) \exp[-j2\pi(ux + vy)]dxdy$$

$$= \int_0^T \left[\int_{-\infty}^{\infty} \int_{-\infty}^{\infty} f[x - x_0(t), y - y_0(t)] \exp[-j2\pi(ux + vy)]dxdy \right] dt \tag{4.54}$$

$$= F(u, v) \int_0^T \exp\{-j2\pi[ux_0(t) + vy_0(t)]\}dt$$

By defining

$$H(u, v) = \int_0^T \exp\{-j2\pi[ux_0(t) + vy_0(t)]\}dt \tag{4.55}$$

Equation (4.54) can be written as

$$G(u, v) = H(u, v)F(u, v) \tag{4.56}$$

If the motion components are given, the transfer function $H(u, v)$ can be obtained from eq. (4.55).

A simple example is discussed below. Suppose that there is only motion along the X direction; that is, $t = T$ and $y_0(t) = 0$. When $x_0(t) = ct/T$, the distance moved by $f(x, y)$ is c,

$$H(u, v) = \int_0^T \exp\left[-j2\pi u \frac{ct}{T} \right]dt = \frac{T}{\pi uc} \sin(\pi uc)\exp[-j\pi uc] \tag{4.57}$$

when n takes an integer value, then H is zero at $u = n/c$.

When $f(x, y)$ is zero or is known to be outside of the region $0 \le x \le L$, the problem caused by $H = 0$ can be avoided and the image can be completely reconstructed from the knowledge of $g(x, y)$ in this region.

Substituting $x_0(t) = ct/T$ into eq. (4.53), suppressing the time invariant variable y, and ignoring the scale parameters give

$$g(x) = \int_0^T f\left[x - \frac{ct}{T}\right]dt = \int_{x-c}^x f(\tau)d\tau \qquad 0 \le x \le L \tag{4.58}$$

Taking the differentiation with respect to x can provide an iterative formula

$$f(x) = g'(x) + f(x - c) \qquad 0 \le x \le L \tag{4.59}$$

Assuming that $L = Kc$, K is an integer, and $p(z)$ is the part of scene moved in $0 \le z < c$ during the capturing period,

$$p(z) = f(z - c) \quad 0 \le z < c \tag{4.60}$$

It can be proven that eq. (4.59) can be written as

$$f(x) = \sum_{k=0}^m g'(x - kc) + p(x - mc) \qquad 0 \le x \le L \tag{4.61}$$

where m is the integer part of x/c. Since $g'(x)$ is known, only $p(x)$ is estimated to obtain $f(x)$. A method to directly estimate $p(x)$ from the blurred image is presented in the following.

First, define

$$\tilde{f}(x) = \sum_{j=0}^m g'(x - jc) \tag{4.62}$$

Substituting eq. (4.62) into eq. (4.63), eq. (4.64) becomes

$$p(x - mc) = f(x) - \tilde{f}(x) \tag{4.63}$$

Note that m ranges from 0 to $K - 1$ as x varies from 0 to L. Thus, p is repeated K times during the evaluation of $f(x)$ from 0 to L. For each $kc \le x < (k + 1)c$, eq. (4.62) is calculated. Adding the results for $k = 0, 1, \ldots, K - 1$ gives

$$p(x) = \frac{1}{K}\sum_{k=0}^{K-1} f(x + kc) - \frac{1}{K}\sum_{k=0}^{K-1} \tilde{f}(x + kc) \tag{4.64}$$

The first sum on the right-hand side of eq. (4.64) is unknown, but it approaches the mean value of $f(\cdot)$ for large values of K. Suppose that the sum is taken as a constant A. Substitute eq. (4.62) into eq. (4.64), then

$$
\begin{aligned}
p(x - mc) &\approx A - \frac{1}{K} \sum_{k=0}^{K-1} \tilde{f}(x + kc - mc) \\
&\approx A - \frac{1}{K} \sum_{k=0}^{K-1} \sum_{j=0}^{k} g'(x + kc - mc - jc) \qquad 0 \le x \le L
\end{aligned}
\tag{4.65}
$$

Combined with eq. (4.62) and eq. (4.63),

$$
f(x) \approx A - \frac{1}{K} \sum_{k=0}^{K-1} \sum_{j=0}^{k} g'[x - mc + (k - j)c] + \sum_{j=0}^{m} g'(x - jc)
\tag{4.66}
$$

Reintroducing the suppressed variable y yields the final result

$$
f(x, y) \approx A - \frac{1}{K} \sum_{k=0}^{K-1} \sum_{j=0}^{k} g'[x - mc + (k - j)c, y] + \sum_{j=0}^{m} g'(x - jc, y)
\tag{4.67}
$$

Example 4.2 Removing the blur caused by uniform linear motion

A restoration example for removing the blur caused by uniform linear motion is presented in Figure 4.8. Figure 4.8(a) is a 256 × 256 image, which is captured when there is uniform linear motion between the camera and the object in a scene. Suppose that the moving distance along the horizontal direction is 1/8 of the image size, which is 32 pixels. This image can be restored using eq. (4.66). When the moving distance is 32, the result is very good as shown in Figure 4.8(b). When the moving distance is 24 and 40, the results are rather poor as shown in Figures 4.8(c) and (d). The poor results are due to the inaccurate estimation of the moving speed. ▣

(a) (b) (c) (d)

Figure 4.8: Removing the blur caused by uniform linear motion.

4.4 Techniques for Constrained Restoration

In constrained restoration, some different techniques are used.

4.4.1 Constrained Restoration

The solution to eq. (4.20) can also be considered from the least square sense, which is to minimize the function of $||Q\hat{f}||^2$, where Q is a linear operator on f, subject to the constraint $\left\|g - H\hat{f}\right\|^2 = \|n\|^2$. This problem can be solved by using the method of Lagrange multipliers. Denote l as the Lagrange multiplier (a constant) and the criterion function can be written as

$$L(\hat{f})||Q\hat{f}||^2 + l\left(\left\|g - H\hat{f}\right\|^2 - \|n\|^2\right) \tag{4.68}$$

An $\hat{f}$, which can minimize eq. (4.68) is to be found. Similar to the process for solving eq. (4.44), by solving eq. (4.68), a **constrained restoration** formula can be obtained as

$$\hat{f} = \left[H^TH + sQ^TQ\right]^{-1} H^Tg \tag{4.69}$$

where $s = 1/l$.

Constrained restoration provides more flexibility in the restoration process, as shown by the two techniques discussed in the following two subsections.

4.4.2 Wiener Filter

A **Wiener filter** is a **least mean squares filter**. It can be derived from eq. (4.69).

Denote $E\{\cdot\}$ as the expectation operation, let R_f and R_n be the correlation matrices of f, and n. R_f and R_n can be defined by the following two equations

$$R_f = E\{ff^T\} \tag{4.70}$$
$$R_n = E\{nn^T\} \tag{4.71}$$

The ij-th element of R_f is given by $E\{f_i f_j\}$, which is the correlation between the i-th and the j-th elements of f. Similarly, the ij-th element of R_n is given by $E\{n_i n_j\}$, which is the correlation between the i-th and the j-th elements of n. Since the elements of f and n are real, R_f and R_n are real symmetric matrices. For most image functions, the correlation between pixels does not extend beyond a distance of 20–30 pixels in the image, so a typical correlation matrix only has a strip of nonzero elements about the main diagonal and has zero elements in the right-upper and left-lower corner regions. Based on the assumption that the correlation between any two pixels is a function of the distance between the pixels but not their position, R_f and R_n can be expressed by block-circulant matrices and diagonalized by the matrix W with the procedure described previously.

$$R_f = WAW^{-1} \tag{4.72}$$

$$R_n = WBW^{-1} \tag{4.73}$$

where the elements in A and B correspond to the Fourier transforms of the correlation elements in R_f and R_n, respectively. The Fourier transforms of these correlation elements are called the power spectrums of $f_e(x, y)$ and $n_e(x, y)$, respectively. In the following discussion, they are denoted $S_f(u, v)$ and $S_n(u, v)$.

Now, defining $Q^T Q = R_f^{-1} R_n$, and substituting it into eq. (4.69) yields

$$\hat{f} = (H^T H + s R_f^{-1} R_n)^{-1} H^T g \tag{4.74}$$

With the help of eq. (4.28), eq. (4.29), eq. (4.72), and eq. (4.73),

$$\hat{f} = (WD^* DW^{-1} + s WA^{-1} BW^{-1})^{-1} WD^* W^{-1} g \tag{4.75}$$

Multiplying both sides of eq. (4.75) by W^{-1} gives

$$W^{-1}\hat{f} = (D^* D + s A^{-1} B)^{-1} D^* W^{-1} g \tag{4.76}$$

Since the matrices inside the parentheses are diagonal, the elements of eq. (4.76) can be written as

$$\hat{F}(u, v) = \left[\frac{1}{H(u, v)} \times \frac{|H(u, v)|^2}{|H(u, v)|^2 + s\left[S_n(u, v)/S_f(u, v)\right]} \right] G(u, v) \tag{4.77}$$

Several cases of eq. (4.77) are discussed as follows:
(1) If $s = 1$, the term inside the brackets is the **Wiener filter**.
(2) If s is a variable, the term inside the brackets is called the **parametric Wiener filter**.
(3) If there is no noise, $S_n(u, v) = 0$, the Wiener filter degrades to the ideal **inverse filter**.

Since s must be adjusted to satisfy eq. (4.69), when $s = 1$, the use of eq. (4.77) no longer yields an optimal solution as defined in the above, since s must be adjusted to satisfy the constraint of $\|g - H\hat{f}\|^2 = \|n\|^2$. However, this solution is optimal in the sense of minimizing $E\{[f(x, y) - \hat{f}(x, y)]^2\}$.

In practice, $S_n(u, v)$ and $S_f(u, v)$ are often unknown, so eq. (4.77) can be approximated by

$$\hat{F}(u, v) \approx \left[\frac{1}{H(u, v)} \times \frac{|H(u, v)|^2}{|H(u, v)|^2 + k} \right] G(u, v) \qquad (4.78)$$

where K is a constant.

Example 4.3 Some comparisons of the inverse filter and the Wiener filter

Some comparisons of the inverse filter and the Wiener filter are shown in Figure 4.9. The column images in Figure 4.9(a) are obtained by convolving the original cameraman image with a smoothing function $h(x, y) = \exp[\sqrt{(x^2 + y^2)}/240]$, then adding a Gaussian noise, whose mean is zero and variance equals to 8, 16, and 32, respectively. The column images in Figure 4.9(b) are obtained by using the inverse filter and the column images in Figure 4.9(c) are obtained by using the Wiener filter. Comparing Figures 4.9(b) and (c), the superiority of the Wiener filter over the inverse filter is evident. This superiority is particularly visible with stronger noise.

(a)   (b)   (c)

Figure 4.9: Some comparisons of the inverse filter and the Wiener filter.

4.4.3 Constrained Least Square Restoration

Wiener filtering is a statistical method. The optimal criterion used by the Wiener filter is based on the correlation matrices of the image and the noise, respectively. It is thus optimal only in an average sense. The following described restoration procedure, called the constrained least square restoration, is optimal for each image. It only requires the knowledge of the noise's mean and variance.

4.4.3.1 Derivation

The **constrained least square restoration** starts also from eq. (4.69), and the key issue is still to determine the transform matrix **Q**. Equation (4.69) is an ill-conditioning equation that sometimes yields solutions that are obscured by large oscillating values. One solution used to reduce oscillation is to formulate a criterion of optimality based on a measure of smoothness. Such criteria can be based on the function of the second derivative. The second derivative of $f(x, y)$ at (x, y) can be approximated by

$$\frac{\partial^2 f}{\partial x^2} + \frac{\partial^2 f}{\partial y^2} \approx 4f(x, y) - [f(x + 1, y) + f(x - 1, y) + f(x, y + 1) + f(x, y - 1)] \qquad (4.79)$$

The second derivative of $f(x, y)$ in eq. (4.79) can be obtained by convolving $f(x, y)$ with the following operator

$$p(x, y) = \begin{bmatrix} 0 & -1 & 0 \\ -1 & 4 & -1 \\ 0 & -1 & 0 \end{bmatrix} \qquad (4.80)$$

One optimal criterion based on the above second derivative is

$$\min \left[\frac{\partial^2 f(x, y)}{\partial x^2} + \frac{\partial^2 f(x, y)}{\partial y^2} \right]^2 \qquad (4.81)$$

To avoid wraparound error in the discrete convolution process, $p(x, y)$ needs to be extended to

$$p_e(x, y) = \begin{cases} p(x, y) & 0 \le x \le 2 \quad\quad \text{and} \quad 0 \le y \le 2 \\ 0 & 3 \le x \le M - 1 \quad \text{or} \quad\quad 3 \le y \le N - 1 \end{cases} \qquad (4.82)$$

The extension of $f(x, y)$ can be referred to eq. (4.16). If the size of $f(x, y)$ is $A \times B$ and the size of $p(x, y)$ is 3×3, the selection of M and N are $M \ge A + 3 - 1$ and $N \ge B + 3 - 1$, respectively.

The above smoothness criterion can be expressed in matrix form. First, a block-circulant matrix is constructed as

$$C = \begin{bmatrix} C_0 & C_{M-1} & \cdots & C_1 \\ C_1 & C_0 & \cdots & C_2 \\ \vdots & \vdots & \ddots & \vdots \\ C_{M-1} & C_{M-2} & \cdots & C_0 \end{bmatrix} \qquad (4.83)$$

where each sub-matrix C_j is an $N \times N$ circulant constructed from the j-th row of $p_e(x, y)$

$$C_j = \begin{bmatrix} p_e(j, 0) & p_e(j, N-1) & \cdots & p_e(j, 1) \\ p_e(j, 1) & p_e(j, 0) & \cdots & p_e(j, 2) \\ \vdots & \vdots & \ddots & \vdots \\ p_e(j, N-1) & p_e(j, N-2) & \cdots & p_e(j, 0) \end{bmatrix} \qquad (4.84)$$

According to the discussion in Section 4.2, C can be diagonalized by the matrix W

$$E = W^{-1}CW \qquad (4.85)$$

where E is a diagonal matrix whose elements are given by

$$E(k, i) = \begin{cases} P(\lfloor k/NN \rfloor, k \bmod N) & \text{if} \quad i = k \\ 0 & \text{if} \quad i \neq k \end{cases} \qquad (4.86)$$

where $P(u, v)$ is the 2-D Fourier transform of $p_e(x, y)$.
   If it is required that the following constraint

$$\|g - H\hat{f}\|^2 = \|n\|^2 \qquad (4.87)$$

should be satisfied, then the optimal solution is given by eq. (4.69) with $Q = C$. This is given by

$$\hat{f} = (H^T H + sC^T C)^{-1} H^T g = (WD^* DW^{-1} + sWE^* EW^{-1})^{-1} WD^* W^{-1} g \qquad (4.88)$$

Multiplying both sides of eq. (4.88) by W^{-1} gives

$$W^{-1}\hat{f} = (D^* D + sE^* E)^{-1} D^* W^{-1} g \qquad (4.89)$$

According to the discussion in Section 4.2, the elements of eq. (4.89) can be expressed in the following form

$$\hat{F}(u, v) = \left[\frac{H^*(u, v)}{|H(u, v)|^2 + s|P(u, v)|^2} \right] G(u, v) \quad u, v = 0, 1, \cdots, M-1 \qquad (4.90)$$

Equation (4.90) resembles the parametric Wiener filter. The main difference is that no explicit knowledge of the statistical parameters other than an estimation of the noise's mean and variance is required.

4.4.3.2 Estimation
Equation (4.69) indicates that s needs to be adjusted to satisfy the constraint $\|g - H\hat{f}\|^2 = \|n\|^2$. In other words, only when s satisfies this condition, is eq. (4.90)

optimal. A procedure for estimating s is introduced below. First, a residual vector r is defined

$$r = g - H\hat{f} = g - H(H^TH + sC^TC)^{-1}H^Tg \tag{4.91}$$

where r is a function of s. It can be proven that $q(s) = r^Tr = ||r||^2$ and $q(s)$ is a monotonically increasing function of s. It is now expected to adjust s to satisfy

$$||r||^2 = ||n||^2 \pm a \tag{4.92}$$

where a is an accuracy factor. If $||r||^2 = ||n||^2$, the constraint $||g - H\hat{f}||^2 = ||n||^2$ will be strictly satisfied. One simple approach to find an s that satisfies eq. (4.92) is to:
(1)   Specify an initial value of s.
(2)   Compute $\hat{f}$ and $||r||^2$.
(3)   Stop if eq. (4.92) is satisfied; otherwise return to step 2 after increasing s if $||r||^2 <$ $||n||^2 - a$ or decreasing s if $||r||^2 > ||n||^2 + a$.

To implement the above procedure for the constrained least square restoration, some knowledge of $||r||^2$ and $||n||^2$ are required. According to the definition of the residual vector r in eq. (4.91), its Fourier transform is

$$R(u, v) = G(u, v) - H(u, v)\hat{F}(u, v) \tag{4.93}$$

Computing the inverse Fourier transform of $R(u, v)$, $r(x, y)$ can be obtained. From $r(x, y)$, $||r||^2$ can be computed

$$||r||^2 = \sum_{x=0}^{M-1}\sum_{y=0}^{N-1} r^2(x, y) \tag{4.94}$$

Now consider the computation of $||n||^2$. The noise variance of the whole image can be estimated using the expected value

$$\sigma_n^2 = \frac{1}{MN}\sum_{x=0}^{M-1}\sum_{y=0}^{N-1} [n(x, y) - m_n]^2 \tag{4.95}$$

where

$$m_n = \frac{1}{MN}\sum_{x=0}^{M-1}\sum_{y=0}^{N-1} n(x, y) \tag{4.96}$$

Similar to eq. (4.95), the two summations in eq. (4.96) are just $||n||^2$, which means

$$||n||^2 = MN[\sigma_n^2 + m_n^2] \tag{4.97}$$

Figure 4.10: Comparisons of the Wiener filter and the constrained least square filter.

Example 4.4 Comparisons of the Wiener filter and the constrained least square filter
Some comparisons of the Wiener filter and the constrained least square restoration are shown in Figure 4.10. The image in Figure 4.10(a) is obtained by filtering the original cameraman image with a blurring filter. The images in Figures 4.10(b) and (c) are the restoration results of Figure 4.10(a), obtained by using the Wiener filter and the constrained least square filter, respectively. The image in Figure 4.10(d) is obtained by adding some random noise to Figure 4.10(a). The images in Figures 4.10(e) and (f) are the restoration results of Figure 4.10(d) obtained by using the Wiener filter and the constrained least square filter, respectively. From Figure 4.10, when an image is degraded only by blurring (no noise), the performances of the Wiener filter and the constrained least square filter are similar. When an image is degraded by both blurring and adding noise, the performance of the constrained least square filter is better than that of the Wiener filter.

4.5 Interactive Restoration

The above discussion is for automatic restoration. In practical applications, the advantage of human intuition can be used. By interactive restoration, the human knowledge is used to control the restoration process and to obtain some special effects. One example is presented below.

One of the simplest image degradations that can be dealt with by **interactive restoration** is the case when a 2-D sinusoidal interference pattern (**coherent noise**) is superimposed on an image. Let $\eta(x, y)$ denote a sinusoidal interference pattern whose amplitude is A and frequency component is denoted (u_0, v_0), that is,

$$\eta(x, y) = A\sin(u_0 x + v_0 y). \tag{4.98}$$

The Fourier transform of $\eta(x, y)$ is

$$N(u, v) = \frac{-jA}{2}\left[\delta\left(u - \frac{u_0}{2\pi}, v - \frac{v_0}{2\pi}\right) - \delta\left(u + \frac{u_0}{2\pi}, v + \frac{v_0}{2\pi}\right)\right] \tag{4.99}$$

The transform has only imaginary components. This indicates that a pair of impulses of strength $-A/2$ and $A/2$ are located at coordinates $(u_0/2\pi, v_0/2\pi)$ and $(-u_0/2\pi, -v_0/2\pi)$ in the frequency plane, respectively. When the degradation is caused only by the additive noise, it has

$$G(u, v) = F(u, v) + N(u, v) \tag{4.100}$$

In the display of the magnitude of $G(u, v)$ in the frequency domain, which contains the sum of $F(u, v)$ and $N(u, v)$, two noise impulses of $N(u, v)$ would appear as bright dots if they are located far enough from the origin and A is large enough. By visually identifying the locations of these two impulses and using an appropriate band-pass filter at the locations, the interference can be removed from the degraded image.

Example 4.5 Removing sinusoidal interference patterns interactively
One example of removing sinusoidal interference patterns interactively is shown in Figure 4.11. Figure 4.11(a) shows the result of covering the original cameraman image by the sinusoidal interference pattern defined in eq. (4.98). The Fourier spectrum magnitude is shown in Figure 4.11(b), in which a pair of impulses appears in the cross points of the bright lines. These two impulses can be removed by interactively putting two band-rejecting filters at the locations of the impulses. Taking the inverse Fourier transform of the filtering result, the restored image can be obtained as shown in Figure 4.11(d). Note that the diameter of the band-rejecting filter should be carefully selected. If two band-rejecting filters as shown in Figure 4.11(e) are used, in which the diameter of the filter is five times larger than that of the one used in Figure 4.11(c), the filtered image will be the one as shown in Figure 4.11(f). The ring effect is clearly visible as shown in Figure 4.11(f). ▢

In real applications, many sinusoidal interference patterns exist. If many band-rejecting filters are used, too much image information will be lost. In this case, the principal frequency of the sinusoidal interference patterns should be extracted. This can be made by placing, at each bright point, a band-pass filter $H(u, v)$. If an $H(u, v)$

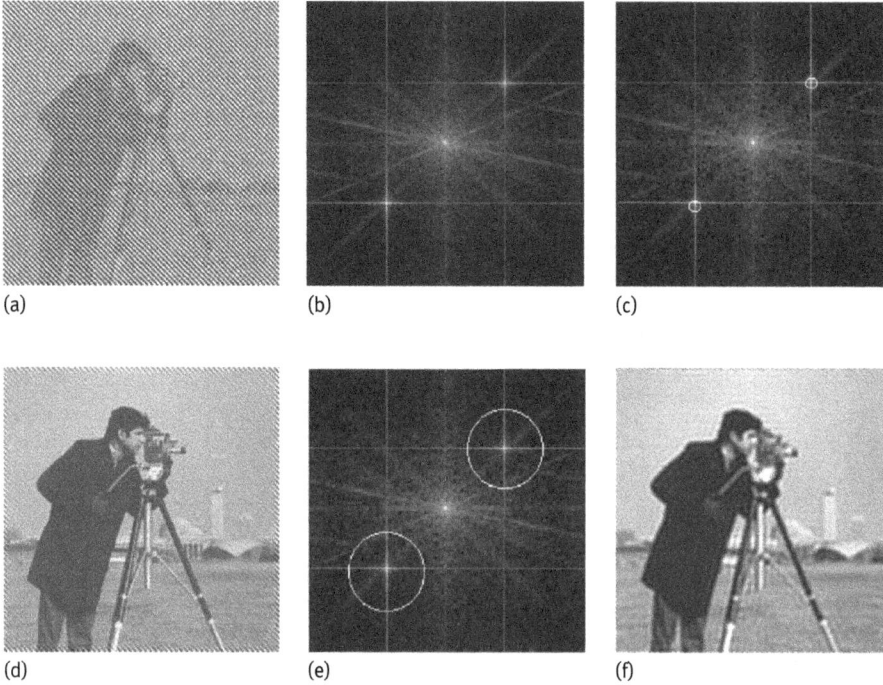

(a)   (b)   (c)

(d)   (e)   (f)

Figure 4.11: Removing sinusoidal interference patterns interactively.

can be constructed, which only permit the components related to the interference pattern to pass, the Fourier transform of this pattern is

$$P(u, v) = H(u, v)G(u, v) \qquad (4.101)$$

The construction of $H(u, v)$ requires judgment of whether a bright point is an interference spike or not. For this reason, the band-pass filter generally is constructed interactively by observing the spectrum of $G(u, v)$. After a particular filter has been selected, the corresponding pattern in this spatial domain is obtained from the expression.

$$p(x, y) = \mathscr{F}^{-1}\{H(u, v)G(u, v)\} \qquad (4.102)$$

If $p(x, y)$ were known completely, subtracting $p(x, y)$ from $g(x, y)$ will give $f(x, y)$. In practice, only an approximation of the true pattern can be obtained. The effects of the components that are not considered in the estimation of $p(x, y)$ can be minimized by subtracting $g(x, y)$ by weighted $p(x, y)$, which is given by

$$\hat{f}(x, y) = g(x, y) - w(x, y)\, p(x, y) \qquad (4.103)$$

where $w(x, y)$ is a weighting function. The weighting function $w(x, y)$ can be selected to get optimal results under certain conditions. For example, one approach is to select

$w(x, y)$ so that the variance of $\hat{f}(x, y)$ is minimized over a specified neighborhood of every point (x, y). Consider a neighborhood of size $(2X + 1) \times (2Y + 1)$ about a point (x, y). The mean and the variance of $\hat{f}(x, y)$ at coordinates (x, y) are

$$\bar{\hat{f}}(x, y) = \frac{1}{(2X + 1)(2Y + 1)} \sum_{m=-X}^{X} \sum_{n=-Y}^{Y} \hat{f}(x + m, y + n) \qquad (4.104)$$

$$\sigma^2(x, y) = \frac{1}{(2X + 1)(2Y + 1)} \sum_{m=-X}^{X} \sum_{n=-Y}^{Y} \left[\hat{f}(x + m, y + n) - \bar{\hat{f}}(x, y) \right]^2 \qquad (4.105)$$

Substituting eq. (4.103) into eq. (4.105) and assuming $w(x, y)$ remains constant over the neighborhood, eq. (4.105) becomes

$$\begin{aligned}
\sigma^2(x, y) &= \frac{1}{(2X + 1)(2Y + 1)} \sum_{m=-X}^{X} \sum_{n=-Y}^{Y} \{[g(x + m, y + n) \\
&\quad - w(x + m, y + n)p(x + m, y + n)] - [g(x, y) - \overline{w(x, y)p(x, y)}]\}^2 \\
&= \frac{1}{(2X + 1)(2Y + 1)} \sum_{m=-X}^{X} \sum_{n=-Y}^{Y} \{[g(x + m, y + n) \\
&\quad - w(x, y)p(x + m, y + n)] - [\bar{g}(x, y) - w(x, y)\bar{p}(x, y)]\}^2
\end{aligned} \qquad (4.106)$$

Solving eq. (4.106), a $w(x, y)$ that minimizes $\sigma^2(x, y)$ is derived as

$$w(x, y) = \frac{\overline{g(x, y)p(x, y)} - \bar{g}(x, y)\bar{p}(x, y)}{\overline{p^2}(x, y) - \bar{p}^2(x, y)} \qquad (4.107)$$

4.6 Image Repairing

In the process of image acquisition, transmission, and processing, some areas of the image may be damaged (with defect) or even missing, such as the sudden change of pixel gray-values, the loss of important parts. Such situations can have many sources and manifestations:

(1) The missing of partial contents in acquiring the image of scene with object occlusion or scanning the old pictures with damaged parts;

(2) The blank left after removing some specific areas (regardless of the scene contents) in the procedure of image operation;

(3) The appearance alteration induced by the text coverage or by the tearing or scratching of picture;

(4) The loss of partial information caused by image lossy compression;

(5) The loss of certain pixels in the (network) transmission of image due to network failure.

Methods and techniques to solve the abovementioned problems are generally called image inpainting and are closely linked to many image restoration problems.

From a technical point of view, these methods for treating the above problems can be further classified into two groups: inpainting and completion. The former comes from the restoration of museum art works on behalf of the oil painting interpolation in the early days. The latter is also often called region completion, in which some missing areas are filled. The top title for these two groups can be called image repairing.

4.6.1 The Principle of Image Repairing

Image repairing is based on the incomplete image and the *a prior* knowledge of the original image, through the use of appropriate methods to correct or rectify the problem of the region defect in order to achieve the restoration of the original image. Repair can be divided into inpainting and completion. Generally, the repair of small-scale area is called **inpainting**, while the repair of large-scale area is called **completion**. Both processes are to restore and complete the damaged or missed information parts in the image, so there are no strict limits on the scale. However, the quantitative change will be caused by qualitative change; the two processes and methods used have their own characteristics from the current technology points of view. Inpainting uses more local structural information of the image rather than the regional texture information, while completion often needs to consider filling the entire image region and makes use of texture information. In terms of function, the former is mostly used for image restoration and the latter for scene object removal and region filling.

It should be pointed out that if there is no prior knowledge of the missing information in the image or the cause of the missing is unknown, the repair of the image is an ill-posed problem, and the solution is uncertain. In practice, the repair of the image often needs to establish a certain model and to make a certain estimate, which are the same for inpainting and completion.

The difficulty or complexity of image repairing (especially completion) comes from three aspects (Chan and Shen, 2005):
(1) Domain complexity: The areas to be repaired will be different with the application domains. For example, in the case of overlay text removal, the area is composed of characters; in the removal of the unwanted object, the area may be of arbitrary shape.
(2) Image complexity: The properties of the image at various scales are different. For example, there are many details/structures at a small scale, while in the large scale the smoothing function approximation can be used.
(3) Pattern complexity: A visually meaningful pattern (high-level meaning) must be considered. Two examples are shown in Figure 4.12. Figure 4.12(a) shows a pair of drawings for filling the center gray patches. When looking at the left drawing,

it is reasonable to use a black block for filling; but when looking at the right drawing, the white block is more reasonable to use. Figure 4.12(b) shows a pair of graphs for filling the missing areas of vertical bars. When looking at the aspect ratio in left graph, the vertical bar seems to belong to the background, then the task should be to distinguish between "E" and "3". When looking at the aspect ratio in right graph, the vertical bar seems to belong to the foreground, then the task should be to recover the partially occluded letter "B".

Image defect as a special case of image degradation, has its own characteristics. After the image is affected by the imperfection, some of the regions may be completely lost, but other regions may not be changed at all in many cases.

Considering an original image $f(x, y)$, its spatial region of pixel distribution is denoted by F. Let the missing part (the part to be repaired) be $d(x, y)$, its spatial region of pixel distribution is denoted by D. For an image $g(x, y)$ to be repaired, its spatial region of pixel distribution is also F, but some parts remain in the original state and others missing. The so-called repair is to use the information from original state region, that is, $F-D$, to estimate and restore the missing information in D.

One illustration example is given in Figure 4.13, where the left image is the original image $f(x, y)$ and the right image is the image to be repaired $g(x, y)$, in which the region D represents the part to be repaired (the original information are completely lost). The region $F-D$ represents the portion of the original image that can be used to repair the region D, also called the source region, and the region D is also called the target region.

With the help of the degradation model shown in eq. (4.8), the model of image repairing can be expressed as

$$[g(x, y)]_{F-D} = \{H[f(x, y)] + n(x, y)\}_{F-D} \tag{4.108}$$

The left side of the equation is the portion of the degraded image that does not degenerate. The goal of image repairing is to estimate and reconstruct $\{f(x, y)\}_D$ by

Figure 4.12: Pattern complexity examples.

Figure 4.13: Indication of different regions in image repairing.

using eq. (4.108). On the one hand, the gray level, color, and texture in the region D should be corresponding to or coordinated with the gray level, color, texture, and so on around the D; on the other hand, the structural information around D should be extended to the interior of D (e. g., the break edges and the contour lines should be connected).

Incidentally, if the noise points in the image are regarded as a target region, the image denoising problem may be treated as an image repairing problem. In other words, the gray scale of the pixel affected by the noise is restored by using the pixels not affected by the noise. If the repair of the image affected by the text overlay or the scratches is regarded as the repair of the curvilinear target area and the repair of the area removed from scene image is regarded as the repair of the planar target area, the repair of the image affected by the noise can be considered as the repair of a patchy target area. The above discussion focuses mainly on impulse noise, because the intensity of impulse noise is very big, the superimposition on the image will make the gray scale of the affected pixel become the extreme value, the original pixel information is covered by the noise completely. In the case of Gaussian noise, pixels with noise superimposed often contain still the original grayscale information, while the pixels in the target area of the image are generally no longer contain the original image information (information are removed).

4.6.2 Image Inpainting with Total Variational Model

The repair techniques used to remove scratches in target areas that are relatively small in size (including linear or curvilinear areas, smaller in one dimension, such as strokes, ropes, and text) are discussed first. The methods commonly used here are based on partial differential equations (PDE) or variational models, both of which can be equivalently introduced by the variational principle. This kind of image repairing methods achieves the purpose of repairing the image by spreading the target area with a way of pixel by pixel. A typical approach is to propagate from the source region to the target region along an equal intensity line (line of equal gray values) (Bertalmio *et al.*, 2001), which helps preserve the structural properties of the image itself. The total variational (TV) model can be used to recover the missing information in the image. A further improvement is the curvature-driven diffusion (CDD) equation, which is proposed for using the continuity principle that does not hold in the total variational model (Chan and Shen, 2001). The advantage of this method is that the linear structure in the image can be well maintained. The disadvantage of this method is that the details of the image are not always preserved, because the blurring may be introduced in the diffusion process, especially when the large target area is repaired.

4.6.2.1 Total Variational Model
The **total variational model** is a basic and typical image repair model. The total variational algorithm is a non-isotropic diffusion algorithm that can be used to denoise while preserving edge continuity and sharpness.

Defining the cost function of diffusion as follows:

$$R[f] = \iint\limits_{F} |\nabla f(x, y)| \, dxdy \tag{4.109}$$

where ∇f is the gradient of f. Consider the case of Gaussian noise, in order to remove noise, eq. (4.109) is also subject to the following constraints

$$\frac{1}{\|F - D\|} \iint\limits_{F-D} |f - g|^2 \, dxdy = \sigma^2 \tag{4.110}$$

where $\|F{-}D\|$ is the area of the region $F{-}D$, and σ is the noise mean square error. The objective of eq. (4.109) is to repair the region and to keep its boundary as smooth as possible, while the function of eq. (4.110) is to make the repairing process robust to noise.

With the help of Lagrangian factor λ, eq. (4.109) and eq. (4.110) can be combined to convert the constrained problem into the unconstrained problem

$$E[f] = \iint\limits_{F} |\nabla f(x, y)| \, dxdy + \frac{\lambda}{2} \iint\limits_{F-D} |f - g|^2 \, dxdy \tag{4.111}$$

If the extended Lagrangian factor λ_D is introduced,

$$\lambda_D(r) = \begin{cases} 0 & r \in D \\ \lambda & r \in (F - D) \end{cases} \tag{4.112}$$

The functional (4.111) becomes

$$J[f] = \iint\limits_{F} |\nabla f(x, y)| \, dxdy + \frac{\lambda_D}{2} \iint\limits_{F} |f - g|^2 \, dxdy \tag{4.113}$$

According to the variational principle, the corresponding energy gradient descent equation is obtained

$$\frac{\partial f}{\partial t} = \nabla \cdot \left[\frac{\nabla f}{|\nabla f|} \right] + \lambda_D(f - g) \tag{4.114}$$

In eq. (4.114), $\nabla \cdot$ represents the divergence.

Equation (4.114) is a nonlinear reaction-diffusion equation with a diffusion coefficient of $1/|\nabla f|$. Within the region D to be repaired, λ_D is zero, so eq. (4.114) degenerates into a pure diffusion equation; while around the region D to be repaired, the second term of eq. (4.114) makes the solution of the equation tending to the original image. The original image can be finally obtained by solving the partial differential equation eq. (4.114).

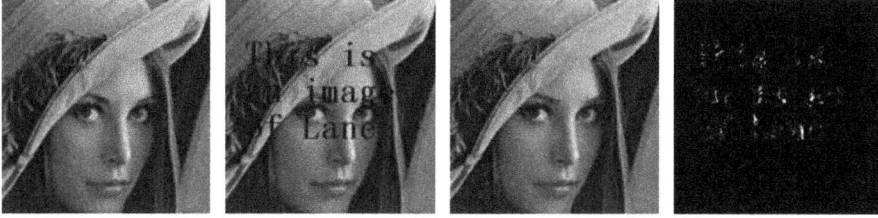

Figure 4.14: Image inpainting for removing overlay text.

Example 4.6 Image inpainting
A set of images in Figure 4.14 gives an example of image inpainting for removing over-lay text. The first three images, from left to right, are the original image, superimposed text image (to be repaired image), and the results of inpainting. The rightmost image is the difference image (equalized by histogram equalization for display) between the original image and the inpainting result image, where the PSNR between the two images is about 20 dB.

4.6.2.2 Hybrid Models

In the total variational model described above, diffusion is only performed in the orthogonal direction of the gradient (*i. e.*, the edge direction), not in the gradient direction. This feature of the total variational model preserves the edge when the diffusion is near the contour of the region; however, in the smooth position within the region, the edge direction will be more random, and the total variation model may produce false contours due to noise influence.

Consider that the gradient term in the cost function is substituted by the gradient square term in the total variational model

$$R[f] = \iint_F |\nabla f(x, y)|^2 \, dxdy \tag{4.115}$$

and in addition, eq. (4.110) is also used for transforming the problem into an uncon-strained problem. Then, by using the extended Lagrangian factor of eq. (4.112), a functional can be obtained

$$J[f] = \iint_F |\nabla f(x, y)|^2 \, dxdy + \frac{\lambda_D}{2} \iint_F |f - g|^2 \, dxdy \tag{4.116}$$

This results in a harmonic model. **Harmonic model** is an isotropic diffusion, in which the edge direction and gradient direction are not be distinguished, so it can reduce the impact of noise generated by false contours, but it may lead to a certain fuzziness for the edge.

A **hybrid model** of weighted sums of two models takes the gradient term in the cost function as

$$R_h [f] = \iint_F h |\nabla f(x, y)| + \frac{(1-h)}{2} |\nabla f(x, y)|^2 \, dxdy \qquad (4.117)$$

where, $h \in [0, 1]$ is the weighting parameter. The functional of the hybrid model is

$$J_h [f] = \iint_F h |\nabla f(x, y)| + \frac{(1-h)}{2} |\nabla f(x, y)|^2 \, dxdy + \frac{\lambda_D}{2} \iint_F |f - g|^2 \, dxdy \qquad (4.118)$$

Comparing eq. (4.109) with eq. (4.118) shows that when $h = 1$, eq. (4.118) is the total variational model.

Another hybrid model combining the two models is the p-harmonic model, where the gradient term in the cost function is

$$R_p [f] = \iint_F |\nabla f(x, y)|^p \, dxdy \qquad (4.119)$$

where, $p \in [1, 2]$ is the control parameter. The functional of the p-harmonic model is

$$J_p [f] = \iint_F |\nabla f(x, y)|^p \, dxdy + \frac{\lambda_D}{2} \iint_F |f - g|^2 \, dxdy \qquad (4.120)$$

Comparing eq. (4.113) with eq. (4.120) shows that when $p = 1$, eq. (4.120) is the total variational model. Comparing eq. (4.116) with eq. (4.120) shows that when $p = 2$, eq. (4.120) is the p-harmonic model. Thus, selecting $1 < p < 2$ could achieve a better balance between the two models.

4.6.3 Image Completion with Sample-Based Approach

The method described in the previous section is more effective for repairing missing regions with small dimensions, but there are some problems when the missing region is large. On the one hand, the method of the previous section is to diffuse the information around the missing region into the missing region. For the larger scale missing region, the diffusion will cause a certain blur, and the blur degree increases with the scale of the missing region. On the other hand, the method of the previous section does not take into account the texture characteristics inside the missing region, and moves the texture characteristics around the missing region directly into the missing region. Because of the large size of the missing regions, the texture characteristics inside and outside may have a greater difference, so the repair results are not ideal.

The basic ideas to solve the above problems include the following two kinds:

(1) Decomposing the image into structural parts and texture parts, the diffusion method described in the previous section can still be used to fill the target region having the strong structure, while for the regions with strong texture the filling will be made with the help of technology of texture synthesis.

(2) Selecting some sample blocks in the nondegraded portion of the image, and using these sample blocks to replace the blocks at the boundary of the region to be filled (the blocks in the nondegraded portions of the image have characteristics close to those boundary blocks). By repeating the filling process progressively toward the inside of the region to be filled to achieve image completion.

The first approach is based on a hybrid approach. Because the natural image is composed of texture and structure, the diffusion method using the structural information is reasonable. However, only using texture synthesis to fill large target region has still some risk and difficulty.

The method based on the second idea is often referred to as **sample-based image completion** method. Such a method directly fills the target region with information from the source region. This idea is enriched by texture filling. It searches in the source region, to find the image blocks most similar to the image block in the target region, and to fill these target blocks by direct substitution.

Compared with the PDE-based diffusion method, the sample-based approach often achieves better results in filling texture content, especially when the scale of the target region is large.

The sample-based image completion method uses the original spatial region to estimate and fill the missing information in the part to be repaired (Criminisi *et al.*, 2003). To this end, a grayscale value is assigned to each pixel in the target region (zero can be used to represent pixel to be filled). In addition, a confidence value is also assigned to each pixel of the target region. The confidence value reflects the degree of confidence in the gray value of the pixel and is no longer changed after the pixel is filled. On the other hand, a temporary priority value is assigned to the image blocks on the filling fronts during the filling process, to determine the order in which the image blocks are filled. The entire filling process is iterative, including three steps.

4.6.3.1 Calculating the Priority of the Image Block

For larger target regions, the process of filling the image block is conducted from the outside to inside. A priority value is calculated for each image block on the filling front, and the filling order for the image blocks is then determined based on the priority value. Considering the need to keep the structure information, this value is relatively large for image blocks that are located on consecutive edges and surrounded by high confidence pixels. In other words, a region with a stronger continuous edge (human is more sensitive to edge information) has the priority to fill, and a region whose more information are known (which is more likely to be more correct) has also the priority to fill.

The priority value of the image block $P(p)$ centered on the boundary point p can be calculated as follows:

$$P(p) = C(p) \cdot D(p) \tag{4.121}$$

In eq. (4.121), the first term $C(p)$ is also called the confidence term, which indicates the proportion of the perfect pixels in the current image block; the second term, $D(p)$ is also called the data item, which is the inner product of the iso-illuminance line at point p (line of equal gray value) and the unit normal vectors. The more numbers the perfect pixels in the current image block, the bigger the first term $C(p)$. The more consistent the normal direction of the current image block with the direction of the iso-illuminance line, the bigger the value of the second term $D(p)$, which makes the algorithm to have higher priority to fill the image blocks along the direction of the iso-illuminance line, so that the repaired image can keep the structure information of the target region well. Initially, $C(p) = 0$ and $D(p) = 1$.

4.6.3.2 Propagating Texture and Structure Information

When the priority values are calculated for all image blocks on all the fronts, the image block with the highest priority can be determined and then it can be filled with the image block data selected from the source region. In selecting the image blocks from the source region, the sum of squared differences of the filled pixels in the two image blocks should be minimized. In this way, the result obtained from such a filling could propagate both the texture and the structure information from the source region to the target region.

4.6.3.3 Updating the Confidence Value

When an image block is filled with new pixel values, the already existing confidence value is updated with the confidence value of the image block where the new pixel is located. This simple update rule helps to measure the relative confidence between image blocks on the frontline.

A schematic for describing the above steps is given in Figure 4.15. Figure 4.15(a) gives an original image, where T represents the target region, both S (S_1 and S_2) represent source regions, and both B (B_1 and B_2) represent the boundaries between the target region and two source regions, respectively. Figure 4.15(b) shows the image block R_p to be filled, which is currently centered on p on B. Figure 4.15(c) shows two candidate image blocks found from the source regions, which can be used for filling. One of the two blocks, R_u, is at the junction of two regions that can be used for filling (similar with the block to be filled), while the other, R_v, is in the first region that can be used for filling (has a larger gap with the block to be filled). Figure 4.15(d) gives the result of the partial filling by using the nearest candidate block (here R_u, but some combinations of

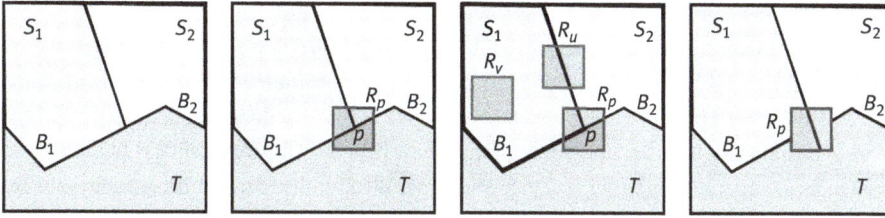

Figure 4.15: Structure propagation schemes based on sample texture filling.

(a) (b) (c)

Figure 4.16: Image completion for removing unwanted object.

multiple candidate image blocks are also possible) to R_p. The above-described filling process can be performed sequentially for all the image blocks in T, and the target region can be finally filled with the contents from source region.

In the filling process, the image blocks can be sequentially selected following the order of peeling onion skin, that is, from outside to inside, ring-by-circle order. If the boundary point of the target region is the center of the image block, the pixels of the image region in the source region are known, and the pixels in the target region are unknown and need to be filled. For this reason, it is also considered to assign weights to the image blocks to be repaired. The more numbers of known pixels are included, the greater the corresponding weights are, and the larger the number of known pixels, the higher the repair order.

Example 4.7 Image completion

A set of images in Figure 4.16 gives an example of removing the (unwanted) object from the image. From left to right are given the original image, the image marking the scope of the object to be removed (image needing to repair) and the image of repair results. Here the object scale is relatively large than text strokes, but the visual effect of repair is relatively satisfactory. ◻

4.7 Problems and Questions

4-1 Make an analytic comparison of the effect of neighborhood averaging for removing Gaussian noise, uniform noise, and impulse noise (with the help of their probability density functions).

4-2* Assume that the model in Figure 4.5 is linear and position-invariant. Show that the power spectrum of the output is $|G(u, v)|^2 = |H(u, v)|^2 |F(u, v)|^2 + |N(u, v)|^2$.

4-3 Consider a linear position invariant image degradation system with impulse response $h(x - s, y - t) = \exp\{-[(x - s)^2 + (y - t)^2]\}$. Suppose that the input to the system is an image consisting of a line of infinitesimal that is located at $x = a$ and modeled by $f(x, y) = \delta(x - a)$. What is the output image $g(x, y)$ if no noise exists?

4-4 Suppose an object has the moving components in a direction that has an angle of $45°$ with respect to the X direction, and the motion is uniform with a velocity of 1. Obtain the corresponding transfer function.

4-5 Suppose that an object has a uniform velocity movement in the X direction for a period of T_x. Then it has a uniform velocity movement in the Y direction for a period of T_Y. Derive the transfer function for these two periods.

4-6 Suppose that the uniform acceleration movement of an object in the X direction causes the blurring of an image. When $t = 0$, the object is in a still state. During the period of $t = 0$ to $t = T$, the acceleration of the object movement is $x_0(t) = at^2/2$. Obtain the transfer function $H(u, v)$ and compare the characteristics of the uniform velocity movement and the uniform acceleration movement.

4-7 Image blurring caused by long-term exposure to atmospheric turbulence can be modeled by the transfer function $H(u, v) = \exp[-(u^2 + v^2)/2\sigma^2]$. Assume the noise can be ignored. What is the equation of the Wiener filter for restoring the blurred image?

4-8* Suppose that a blurring degradation of an imaging system can be modeled by $h(r) = [(r^2 - 2\sigma^2)/\sigma^4] \exp[-r^2/2\sigma^2]$, where $r^2 = x^2 + y^2$. Find out the transfer function of a constrained least square filter.

4-9 Using the transfer function in the above problem, give the expression for a Wiener filter. Assume that the ratio of the power spectra of the noise and the undegraded signal is a constant.

4-10 Suppose that an image is superimposed by a 2-D sinusoidal interference pattern (with a horizontal frequency of 150 Hz and a vertical frequency of 100 Hz).

    (1) What is the difference between the degradation caused by this interference and the degradation caused by the uniform velocity movement?

    (2) Write out the expression for the Butterworth band-pass filter of order 1 (with a cutoff frequency of 50 Hz), which complements the band-rejecting filter for eliminating this interference.

4-11 In many concrete image restoration applications, the parameters used in restoration techniques should be adapted to the *a priori* knowledge of the application. Search in the literature and find what types of *a priori* knowledge can be used.

4-12 (1) For the two images shown in Figure Problem 4-12, discuss on which image is relatively easy to repair and to achieve good result without leaving traces of the effect, and why?

(2) Try to repair them by running an algorithm, and compare the results.

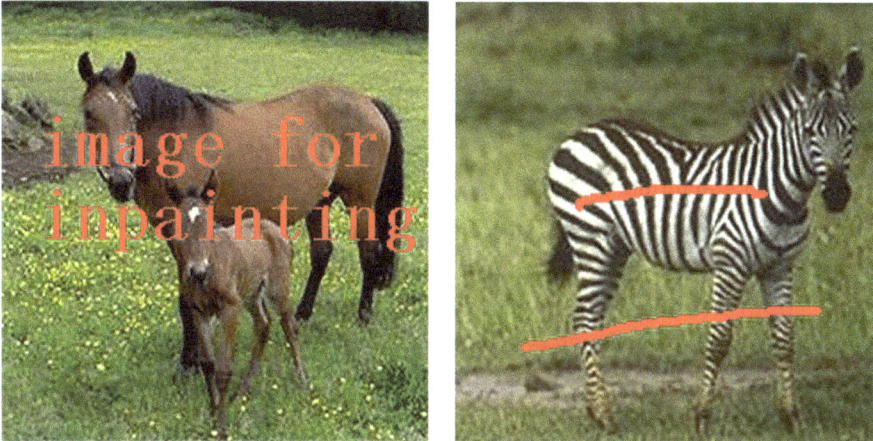

Figure Problem 4-12

4.8 Further Reading

1. **Degradation and Noise**
   - Further discussions on noise can be found in Libbey (1994). Particular descriptions on the probability density functions of Rayleigh noise, Gamma noise, and exponential noise can be found in Gonzalez and Woods (2008).
   - Discussion on the degradations related to the optical systems during the image capturing process can be found in Pratt (2001).
   - One method for noise estimation in an image can be found in Olsen (1993).
   - Impulse noise can be removed parallely (Duan and Zhang, 2011).

2. **Degradation Model and Restoration Computation**
   - Further discussion on various properties of matrices can be found in Committee (1994) and Chen and Chen (2001).
   - More explanation for image restoration can be found in several textbooks; for example, Gonzalez and Woods (2008).
   - Image blurring is one type of image degradation process. A technique based on neural networks to identify and restore a blurred image can be found in Karnaukhov *et al.* (2002).

3. **Techniques for Unconstrained Restoration**
   – Using a sequence of low-resolution images to reconstruct a high-resolution image is called super-resolution reconstruction, which is also an image restoration technique (Tekalp, 1995).

4. **Techniques for Constrained Restoration**
   – More discussion on constrained restoration can be found in Castleman (1996) and Gonzalez and Woods (2008).

5. **Interactive Restoration**
   – Details for deriving related formulas can be found, for example, in Bracewell (1995).

6. **Image Repairing**
   – A general introduction for image repairing can be found in Zhang (2015b).

   – Using sparse representation for image repairing can be found in Shen *et al.* (2009).

   – Using nonnegative matrix factorization and sparse representation for image repairing can be found in Wang and Zhang (2011).

5 Image Reconstruction from Projections

Image reconstruction from projections is a special kind of image processing, whose input is a sequence of projected images and whose output is the reconstructed image. Reconstruction from projections can also be considered as a special kind of image restoration process. If the projection process is a degradation process, the reconstruction process is then a restoration process. During the projection phase, the information along the projection direction has been lost (only the 1-D information is left), and the reconstruction recovers this information from multiple projections.

The sections of this chapter are arranged as follows:

Section 5.1 introduces several typical projection methods and the principle of 2-D and 3-D image reconstruction from projection.

Section 5.2 describes the method of reconstruction by inverse Fourier transform that needs less computation but has relatively poor quality of reconstructed image.

Section 5.3 presents the methods of convolution and back-projection reconstruction, which are easy to implement in software and hardware, and the reconstructed image is clear and accurate.

Section 5.4 introduces the algebra reconstruction technology, which is opposite to transform reconstruction technique, and can obtain the numerical solution directly through iterative calculation.

Section 5.5 provides some discussions on combining the transformation method and the series expansion method (algebra reconstruction).

5.1 Modes and Principles

The problem of image reconstruction from projections has arisen independently from a large number of scientific fields. An important version of a problem in medicine is to obtain the internal structure of the human body from multiple projections recorded using X-rays, γ-rays, neutrons, ultrasounds, etc. This process is referred to as computerized tomography (CT). In Greek, "tomos" means cutting. Tomography is a slice-based representation of a solid object understudy.

5.1.1 Various Modes of Reconstruction from Projections

Nowadays, there are a number of ways to make an internal structure explicitly visible with image reconstruction from projections. Some of these ways are described in the following.

DOI 10.1515/9783110524116-005

5.1.1.1 Transmission Computed Tomography

Transmission computed tomography (TCT) is the most popular use of CT. Since x-rays are often used in TCT, TCT is also called XCT. The radiation emitted from the emission source passes through the object and arrives on the receiver. When the radiations pass through the object, part of them is absorbed by the object, but the rest is received by the receiver. Since different parts of objects absorb different quantities of radiation, the radiation intensity received by the receiver shows the degree of radiation absorption at different parts of an object.

Let I_0 denote the intensity of the source, $k(x)$ denote the linear attenuation factor, L denote the radiation line, and I denote the intensity passing through an object. These parameters have the relation given by

$$I = I_0 \exp\left\{-\int_L k(s)ds\right\} \tag{5.1}$$

If the object is uniform/homogeneous,

$$I = I_0 \exp\{-kL\} \tag{5.2}$$

where I represents the intensity after passing through the object, I_0 represents the intensity without the object, L is the length of radial inside the object, and k is the linear attenuation factor of the object.

In recent years, the scan time needed for CT has decreased dramatically, and the spatial resolution of CT has increased continuously. In 1972, the scan time required by CT was about 5 minutes and the spatial resolution was only 80×80. At present, the scan time for obtaining a 1024×1024 image is less than 1 second.

An ultrasound CT has the same working principles as TCT, except that the emission is ultrasound. Since ultrasound is not transmitted precisely along a line, some nonlinear models are required.

5.1.1.2 Emission Computed Tomography

The history of **emission computed tomography** (ECT) can be traced back to the 1950s. The source of the ECT systems is placed inside the object that is to be examined. A common procedure is to inject ions with radioactivity into the object and detect the emission outside of the object. In this way, the distribution and movement of the ions can be detected, so the physiological states of objects can be investigated. Two kinds of ECT can be distinguished: **positron emission tomography** (PET) and **single positron emission CT** (SPECT). Both provide the space distribution information of radioisotopes. They differ in the methods used to define the ray direction.

PET has great sensitivity to detect details in nanomolar scale. Figure 5.1 illustrates the structure of a PET system. It uses ions with radioactivity, which can emit positrons in attenuation. The emitted (positive) positrons quickly collide with the (negative) electrons and are annihilated; thus, a pair of photons is produced and emitted in

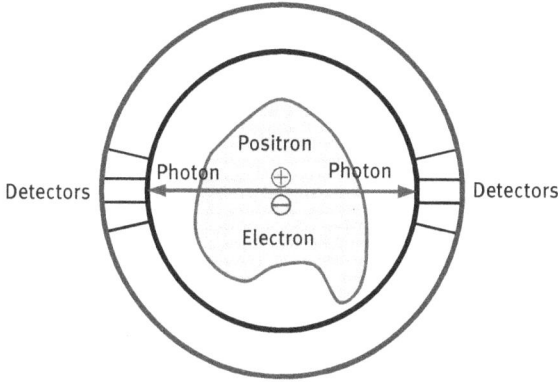

Figure 5.1: The structure of a PET system.

opposite directions. Two detectors installed face-to-face can receive these two photons and determine the radial. In PET systems, all detectors are spread clockwise and any two detectors facing each other form a pair of detectors used to detect the photon pair.

If two photons have been recorded simultaneously by a pair of detectors, then annihilation must happen in the line connecting these two detectors. A recording from such a projection event can be written as

$$P = \exp\left(-\int k(s)ds\right)\int f(s)ds \tag{5.3}$$

where P denotes the projection data, $k(s)$ is the attenuation coefficient, and $f(s)$ is the distribution function of radioisotopes.

SPECT is a technique that combines radiology and tomography. In SPECT, any radioisotope that emits decay γ-ray may be used as the basis for imaging. In contrast to the annihilation radiation, these γ-rays are emitted as single individual photons. It thus allows a much broader range of isotopes to be used, which are easily available. In contrast, because of the short half-lives of the isotopes involved in PET, a very expensive cyclotron is required.

The structure of a SPECT system is shown in Figure 5.2. To determine the ray directions, materials that are opaque to the γ-rays are used to make spatial collimators. These collimators eliminate most of the γ-rays that would otherwise strike the detector and allow only those that are incident in the prescribed direction to pass. The γ-rays arrived at the scintillation crystal are transformed to low energy photons and further transformed to electrical signals by a photomultiplier.

The sensitivity of SPECT is much less than that of PET, because only 0.1% of the γ-rays can pass the collimators. The sensitivity of ECT is

$$S \propto \frac{Ae^n k}{4\pi r^2} \tag{5.4}$$

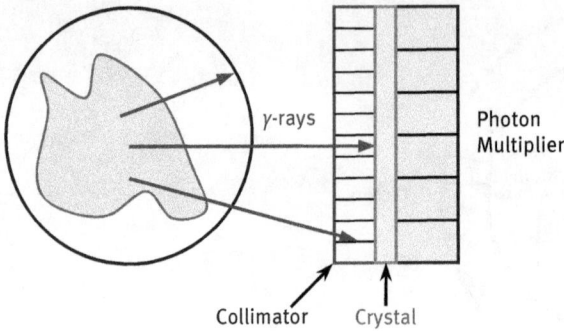

Figure 5.2: The structure of a SPECT system.

where A denotes the area of the detector, e is the efficiency of the detector (1 for SPECT and 2 for PET), k is the attenuation coefficient (around 0.2–0.6), and r is the radius of the object. The sensitivity ratio of PET to SPECT is about 150 divided by the resolution. When the resolution is 7.5 mm, this ratio is about 20.

5.1.1.3 Reflection Computed Tomography

Reflection computed tomography (RCT) also works based on the principle of reconstruction from projections. A typical example is the radar system, in which the radar map is produced by the reflection from objects.

The structure of the synthetic aperture radar (SAR) system is shown in Figure 5.3. In SAR imaging, the radar moves while objects are assumed to be static (their relative movement is used to improve the transverse resolution). Suppose that v is the moving speed of the radar along the Y-axis, T is the effective cumulative time, and λ is the wavelength. Two point-objects are located along the moving direction of the radar, object A is located at the beam-line (X-axis), and the offset between object A and object B is d.

In Figure 5.3, the minimum distance between object A and the radar is R, and this moment is defined as the origin $t = 0$. Suppose that around $t = 0$ the change of distance is δR, then it has $\delta R = (y - d)^2/2R$ for $R \gg \delta R$. The advanced phase of the reflected signal at object A for the round trip is

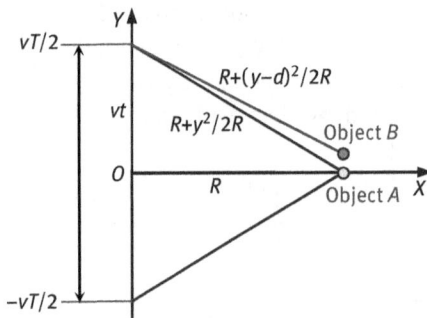

Figure 5.3: The structure of a SAR system.

$$\theta_A(t) = -\frac{4\pi y^2}{2R\lambda} = -\frac{4\pi}{\lambda}\frac{v^2 t^2}{2R} \tag{5.5}$$

The advanced phase of the reflected signal at object B for the round trip is

$$\theta_B(t, d) = -\frac{4\pi}{\lambda}\frac{(vt - d)^2}{2R} \tag{5.6}$$

If the frequency of the emitted signal is high enough, the reflected signal can be considered continuous. Therefore, these signals can be integrated. In addition, if the emission is uniform during the integration period, the reflected signal at object B is

$$E(d) = \int_{-T/2}^{T/2} \exp\left[-\frac{j4\pi}{2R\lambda}(vt - d)^2\right] dt \tag{5.7}$$

5.1.1.4 Magnetic Resonance Imaging

Magnetic resonance imaging (MRI) was originally called nuclear magnetic resonance (NMR). Its process is briefly described below. Materials with an odd number of protons or neutrons possess a weak but observable nuclear magnetic moment. The static nuclear moment is far too weak to be measured when it is aligned with the strong static magnetic field. Resonance techniques permit this weak moment to be measured. The main idea is to measure the moment while it oscillates in a plane perpendicular to the static field.

When protons are placed in a magnetic field, they oscillate (resonate) at a frequency that depends on the field strength and absorb energy at the oscillation frequency. This energy is reradiated as the protons return to their round state. The reradiation involves processes (relaxation of the magnetization components parallel and perpendicular to the field) with different time constants T1 and T2. The strength of the MRI signal depends on the proton concentration (essentially the water concentration in the tissue for medical imaging), but its contrast depends on T1 and T2, which are strongly influenced by the fluid viscosity or tissue rigidity. Weighting the combination of the two signals provides control over the observed image.

The observed images are reconstructed by the integration of resonance signals along different directions in a 3-D space V:

$$S(t) = \iiint_V R(x, y, z)f(x, y, z)\exp[j\theta \int_0^t w(x, y, z, \tau)d\tau]\,dxdydz \tag{5.8}$$

where $R(x, y, z)$ depends on the nucleus of the material; $f(x, y, z)$ is a distribution function; $w(x, y, z, t)$ is a Larmor frequency function, which is written as $w(x, y, z, t) = g[B_0 + B(x, y, z, t)]$ (B_0 is the intensity of magnetic field and $B(x, y, z, t)$ is a time-varying inhomogeneous magnetic field); and $g(\bullet)$ is also related to the nucleus of the material.

MRI is used to recover the intensity distribution function based on the resonance signals inspired by magnetic fields. From a mathematical point of view, MRI is an inverse problem; that is, given $S(t)$, $w(x, y, z, t)$, and $f(x, y, z)$, solve eq. (5.8) for $R(x, y, z)$.

Another MRI method is to represent the received magnetic signals as the function of the density of spins for an object. For an object with the density of spins $R(r)$, whose spatial-frequency signal $S(q)$ is

$$S(q) = \int_{r_1}^{r_2} R(r) \exp[-jq(r) \cdot r] dr \tag{5.9}$$

where the integration covers the whole object. $S(q)$ and $R(r)$ form a pair of Fourier transforms.

5.1.1.5 Electrical Impedance Tomography

Electrical impedance tomography (EIT) is a relatively new medical imaging modality currently under development. It uses noninvasive signals to probe the object and then detects the responses on the boundary of the object in order to reconstruct an impedance distribution inside the object. It is harmless, relatively low in cost, easier to operate, and uses nonionizing radiation. EIT is quite sensitive to the change in conductance or reactance, and is the only method for acquiring the image of the conductance. By injecting low-frequency current into an object and measuring the electrical potential from the outside surface of the object, the image representing the distribution or variation of the conductance and the reactance inside the object can be reconstructed using the techniques of reconstruction from projections. Using this technique, the structure and the inner change of the object materials can also be obtained.

The distribution of the electrical potential within an isotropic conducting object through which a low-frequency current is flowing is given by Barber (2000)

$$\nabla (C\nabla p) = 0 \tag{5.10}$$

where p is the potential distribution within the object and C is the distribution of conductivity within the object. If the conductivity is uniform, it reduces to a Laplace equation. Strictly speaking, this equation is only correct for the direct current. For the frequencies of the current used in EIT (up to 1 MHz) and the sizes of objects being imaged, it can be assumed that this equation continues to describe the instantaneous distribution of the potential within the conducting object. If this equation is solved for a given conductivity distribution and a current distribution through the surface of the object, the potential distribution developed on the surface of the object may be determined. The distribution of the potential depends on several things. It depends on the pattern of the applied current and the shape of the object. It also depends on the internal conductivity of the object, which is to be determined. In theory, the current may be applied in a continuous and nonuniform pattern at every point across the surface. In practice, the current is applied to an object through electrodes attached to the surface of the object. Theoretically, the potential may be measured at every point on the surface of the object. Again, the voltage on the surface of the object is measured in practice using electrodes attached to the surface of the object. There is a relationship between an applied current pattern P_i, p and C

$$p_i = R(P_i, C) \tag{5.11}$$

For one current pattern P_i, the knowledge of p_i is not, in general, sufficient to determine C uniquely. However, by applying a complete set of independent current patterns, it is possible to obtain sufficient information to determine C, at least in the isotropic case. This is the inverse solution.

In practice, measurements of the surface potential or voltage can only be made at a finite number of positions, corresponding to electrodes placed on the surface of the object. This also means that only a finite number of independent current patterns can be applied. For N electrodes, $N-1$ independent current patterns can be defined and $N(N-1)/2$ independent measurements can be made. The latter number determines the limitation of the image resolution achieved with N electrodes. In practice, it may not be possible to collect all possible independent measurements. Since only a finite number of current patterns and measurements are available, eq. (5.11) can be rewritten as

$$V = A_C C \tag{5.12}$$

where V is now a concatenated vector of all voltage values for all current patterns, C is a vector of the conductivity values representing the conductivity distribution divided into uniform image pixels, and A_C is a matrix representing the transformation of this conductivity vector into the voltage vector. Since A_C depends on the conductivity distribution, this equation is nonlinear. Although formally the preceding equation can be solved for C by inverting A_C, the nonlinear nature of this equation means that this cannot be done in a single step. An iterative procedure will therefore be needed to obtain C.

A basic algorithm for reconstruction is as follows. For a given set of the current patterns, a forward transform is set up to determine the voltages V produced from the conductivity distribution C, as shown in eq. (5.12). A_C is dependent on C, so it is necessary to assume an initial conductivity distribution C_0. This distribution is usually assumed to be uniform. Using A_C, the expected voltages V_0 are calculated and compared with the actual measured voltages V_m. It can be proven that an improved estimation of C is given by

$$\Delta C = \left(S_C^T S_C\right)^{-1} S_C^T (V_0 - V_m) \tag{5.13}$$
$$C_1 = C_0 + \Delta C \tag{5.14}$$

where S_C is the differential of A_C with respect to the sensitivity matrix C. The improved value of C is then used in the next iteration to compute an improved estimate of V_m (i. e., V_1). This iterative process is continued until some appropriate endpoint is reached. Although the convergence is not guaranteed, in practice, the convergence to the correct C in the absence of noise can be expected, if a good initial value is chosen. A uniform conductivity seems to be a reasonable choice. In the presence of noise on the measurements, the iteration is stopped when the difference between V and V_m is within a margin set by the known noise.

Currently used EIT systems can be divided into two classes: **applied current electrical impedance tomography** (ACEIT) and **induced current electrical impedance tomography** (ICEIT). The former measures the distribution of the impedance using injection-excitation techniques. It connects the source of the signal excitation to the driving electrodes on the surface of the object, and from the magnitude and phase of the voltage or the current on the measuring electrodes, the impedance in the imaging regions can be determined. Owing to different characteristics of various materials, a single frequency of the voltage or the current is not enough to measure the impedances of different materials. One solution for this problem is to use the sweep frequency technique, which is a relatively new branch of EIT. EIT creates alternative excitation using loops that do not touch the surface of objects. The loop produces the Faradic vortex inside the object, which can be detected from the outside.

Example 5.1 EIT images obtained with ACEIT
Two images reconstructed with ACEIT are shown in Figure 5.4, in which the gray levels represent the value of the impedances.   ▣

From the mathematical point of view, EIT is similar to various CT techniques, as all of them require the use of projection data to recover the inside structure. Compared to CT and MRI, EIT has advantages of real-time techniques, costs less, and is easy to implement for both continuous monitoring and functional imaging.

In summary, if the measured data contain the integration form of some interesting physical properties of an object, the techniques of reconstruction from projections could be used to obtain the images representing those physical properties of the object (Kak and Slaney, 2001).

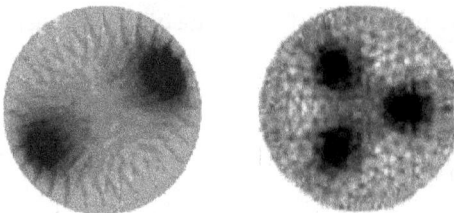

Figure 5.4: EIT images obtained with ACEIT.

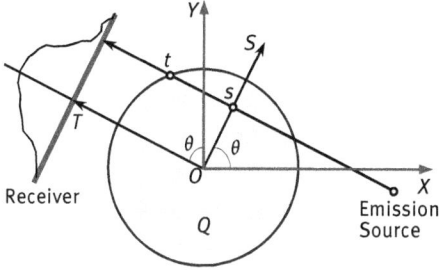

Figure 5.5: Projection of a 2-D function.

5.1.2 The Principle of Reconstruction from Projections

A basic model of reconstruction from projections is shown in Figure 5.5. A set of projections is taken from a 2-D function $f(x, y)$, which represents the distribution of some physical quantities in the 2-D plane.

It is convenient to assume that the values of $f(x, y)$ are zero if (x, y) is outside a unit circle centered at the origin. Suppose a line from the emission source to the receiver crosses (x, y) inside Q on the plane. This line can be described with two parameters: its distance to the origin, denoted s, and the angle between this line and the Y-axis, denoted θ. The integration of $f(x, y)$ along line (s, θ) is

$$g(s, \theta) = \int_{(s,\theta)} f(x, y)dt = \int_{(s,\theta)} f(s \times \cos\theta - t \times \sin\theta, s \times \sin\theta + t \times \cos\theta)dt \qquad (5.15)$$

This integration is the projection of $f(x, y)$ along the direction t, in which the limits depend on s, θ, and Q. When Q is a unit circle and the lower and upper limits of the integration are t and $-t$, respectively,

$$t(s) = \sqrt{1 - s^2} \qquad |s| \leq 1 \qquad (5.16)$$

If the line represented by (s, θ) is outside of Q,

$$g(s, \theta) = 0 \qquad |s| > 1 \qquad (5.17)$$

In real reconstruction from projections, $f(x, y)$ represents the object to be reconstructed, (s, θ) determines an integration path, which corresponds to a line from the emission source to the receiver, and $g(s, \theta)$ denotes the integration result. Under these definitions, reconstruction from projections can be described as the determination of $f(x, y)$, for a given $g(s, \theta)$. Mathematically, this task is to solve the integration in eq. (5.15).

The utilization of tomographic reconstruction makes true 3-D imaging possible. Each individual projection provides a 2-D array, and all projections together form a 3-D array of cubic voxels. In contrast to fan-beam geometry, which is used in

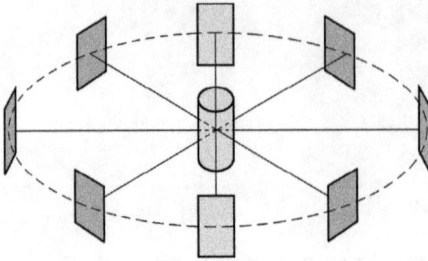

Figure 5.6: 3-D imaging by rotating the sample about a single axis.

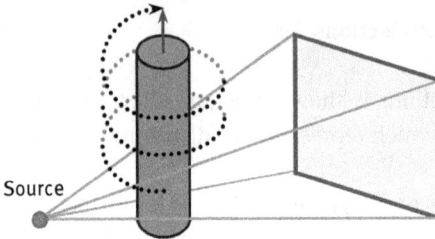

Source

Figure 5.7: 3-D imaging by helical scanning.

single-slice projection, **cone-beam geometry** can be employed here. The set of view directions must include moving out of the plane and going into the 3-D space, which are described by two polar angles. This does not necessarily require rotating the object with two different polar angles, since cone-beam imaging geometry provides different angles for the projection lines, just as a fan beam geometry does in 2-D. The best reconstructions, however, are obtained with a series of view angles that cover the 3-D orientations as uniformly as possible.

A simple geometry method is to rotate the sample around a single axis as shown in Figure 5.6. This method offers the advantage of precise rotation, since the quality of the reconstruction depends on the consistency of the center of rotation. On the other hand, artifacts in the reconstructed voxels can be significant, especially in the direction parallel to the axis and near the north and south poles of the sample. The single-axis rotation method is most often used with X-ray, neutron, or gamma ray tomography because the samples may be rather large and are relatively equiaxed so that the distance through which the radiation must pass is the same in each direction.

Improved resolution in the axial direction can be obtained by using a helical scan (Wang *et al.*, 1991) in which the specimen rotates while it moves in the axial direction as shown in Figure 5.7. This geometry is particularly suitable for small industrial objects.

5.2 Reconstruction by Fourier Inversion

Fourier inversion is a typical reconstruction method among the group of transform-based methods.

5.2.1 Fundamentals of Fourier Inversion

Transform methods of reconstruction from projections consist of three steps:

(1) Establish a mathematical model, in which the unknown quantities and the known quantities are functions, whose arguments come from a continuum of real numbers.

(2) Solve the unknown functions using the inversion formula.

(3) Adapt the inversion formula for the applications with discrete and noisy data.

In step 2, it is found that there are several formulas that have theoretically equivalent solutions to the problem posed in step 1. When each of these formulas is discretized in step 3, it is found that the resulting algorithms do not perform identically on the real data, due to the approximation error introduced in this step.

In the following discussion, the case in which both s and θ are sampled uniformly is considered. Suppose the projections are measured at N angles that are apart from each other by $\Delta\theta$, and M rays that are apart from each other by Δs for each of the angles. Define integers M^+ and M^- as

$$\left.\begin{array}{l} M^+ = (M-1)/2 \\ M^- = -(M-1)/2 \end{array}\right\} \quad M \text{ odd}$$
$$\left.\begin{array}{l} M^+ = (M/2)-1 \\ M^- = -M/2 \end{array}\right\} \quad M \text{ even} \tag{5.18}$$

To cover the unit circle by a set of rays $\{(m\Delta s, n\Delta\theta): M^- \le m \le M^+, 1 \le n \le N\}$, it is needed to choose $\Delta\theta = \pi/N$ and $\Delta s = 1/M^+$. The resulting $g(m\Delta s, n\Delta\theta)$ are referred to as parallel-ray data. In the image domain, a Cartesian grid of the sample points are specified by $\{(k\Delta x, l\Delta y): K^- \le k \le K^+, L^- \le l \le L^+\}$, where K^- and K^+ as well as L^- and L^+ are defined in terms of K and L analogous to eq. (5.18). According to these definitions, a reconstruction algorithm is required to produce $f(k\Delta x, l\Delta y)$ at these $K \times L$ sample points from the $M \times N$ measurements $g(m\Delta s, n\Delta\theta)$.

The basis of transform methods is the **projection theorem for Fourier transform**. Denote $G(R, \theta)$ as the 1-D Fourier transform of $g(s, \theta)$ with respect to the first variable s, that is,

$$G(R, \theta) = \int\limits_{(s,\theta)} g(s, \theta) \exp\left[-j2\pi Rs\right] ds \tag{5.19}$$

$F(X, Y)$ is the 2-D Fourier transform of $f(x, y)$

$$F(X, Y) = \iint\limits_{Q} f(x, y) \exp\left[-j2\pi(xX + yY)\right] dxdy \tag{5.20}$$

Then the following projection theorem can be proven

$$G(R, \theta) = F(R \cos \theta, R \sin \theta) \tag{5.21}$$

which says that the Fourier transform of $f(x, y)$ projected with an angle θ equals the value of the Fourier transform of $f(x, y)$ at (R, θ) in the Fourier space.

5.2.2 Reconstruction Formulas of the Fourier Inverse Transform

Following the projection theorem of the Fourier transform, the reconstruction formulas of the Fourier inverse transform can be easily obtained using a typical transform method for reconstruction from projections. This is given by

$$f(x, y) = \int_{-\infty}^{\infty} \int_{-\infty}^{\infty} G\left[\sqrt{X^2 + Y^2}, \arctan\left(\frac{Y}{X}\right) \right] \exp\left[j2\pi(xX + yY) \right] dX dY \tag{5.22}$$

In a real application, a window is required to limit the integration to a bounded region in the Fourier plane. Thus, a band-limited approximation of $f(x, y)$ can be obtained

$$f_W(x, y) = \iint_Q G\left[\sqrt{X^2 + Y^2}, \arctan\left(\frac{Y}{X}\right) \right] W\left[\sqrt{X^2 + Y^2} \right] \exp\left[j2\pi(xX + yY) \right] dX dY \tag{5.23}$$

To compute $f_W(x, y)$, it is necessary to evaluate $G(\cdot)$, which takes only values for $\theta = \theta_n$ (θ_n denotes $n\Delta\theta$). $G(R, \theta_n)$ can be computed by taking the sum of $g(\bullet)$ at a set of sample points $(m\Delta s, \theta_n)$ as

$$G_\Sigma(R, \theta_n) = \Delta s \sum_{m=M^-}^{M^+} g(m\Delta s, \theta_n) \exp\left[-j2\pi R(m\Delta s) \right] \tag{5.24}$$

If assuming $R = k\Delta R$ (k is an integer, ΔR is the sampling distance), and choose $\Delta R = 1/(M\Delta s)$, this gives

$$G_\Sigma(k\Delta R, \theta_n) = \Delta s \sum_{m=M^-}^{M^+} g(m\Delta s, \theta_n) \exp\left[\frac{-j2\pi km}{M} \right] \tag{5.25}$$

For the arbitrary (X, Y), $G[(X^2 + Y^2)^{1/2}, \arctan(Y/X)]$ can be interpolated by $G_\Sigma(k\Delta R, \theta_n)$. From eq. (5.23), it has

$$G\left[\sqrt{X^2 + Y^2}, \arctan\left(\frac{Y}{X}\right) \right] W\left[\sqrt{X^2 + Y^2} \right] = F_W(X, Y) \tag{5.26}$$

Therefore, $f_W(x, y)$ can be determined from

$$f_W(k\Delta x, l\Delta y) \approx \Delta X \Delta Y \sum_{u=U^-}^{U^+} \sum_{v=V^-}^{V^+} F_W(u\Delta x, v\Delta y)$$

$$\exp\{j2\pi[(k\Delta x)(u\Delta X) + (l\Delta y)(v\Delta Y)]\} \tag{5.27}$$

In addition, let $\Delta x = 1/(U\Delta X)$ and $\Delta y = 1/(V\Delta Y)$, then

$$f_W(k\Delta x, l\Delta y) \approx \Delta X \Delta Y \sum_{u=U^-}^{U^+} \sum_{v=V^-}^{V^+} F_W(u\Delta x, v\Delta y) \exp\left\{j2\pi\left[\frac{ku}{U} + \frac{lv}{V}\right]\right\} \tag{5.28}$$

On the right side of Figure 5.8(a), the locations of $(k\Delta R, \theta_n)$, marked by "•", are drawn for the Fourier plane. At these locations, G can be computed by using eq. (5.25). These points form a polar pattern of samples in the plane, characterized by a uniform sample spacing in the polar coordinates. On the left side of Figure 5.8(a), "+" indicates location $(u\Delta X, v\Delta Y)$, at which the Fourier transform is used to compute $f_W(u\Delta X, v\Delta Y)$ according to eq. (5.28). These points form a Cartesian lattice of samples in the plane. Some interpolation is needed to estimate the values at these sample points from the known values on the polar grid. Equation (5.28) only provides $f_W(k\Delta x, l\Delta y)$ at limited points (the number of k is U and the number of l is V), which can be computed using FFT.

The polar pattern on the right side of Figure 5.8(a) is not efficient for the sampling in the Fourier plane. An improvement is to use the polar pattern shown in Figure 5.8(b). On the right side of Figure 5.8(b), the modified polar patterns have a radial offset of the samples at alternate angles. The distribution of these points is more uniform and is closer to the distribution of the points at the Cartesian lattice on the left side of Figure 5.8(b).

In summary, the reconstruction algorithm based on the Fourier inversion formula has three major subprocesses:
(1) Perform 1-D transforms of the projections at angle $\theta_n(n = 1, 2, \dots, N)$, as in eq. (5.25).

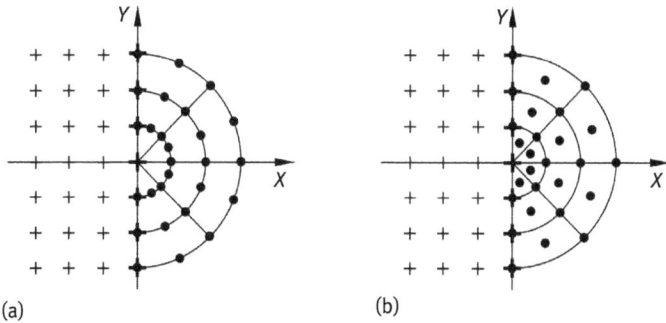

(a)   (b)

Figure 5.8: Cartesian and polar sampling grids in the Fourier space.

(2)   In the Fourier space, interpolate from a polar pattern to a Cartesian lattice.

(3)   Perform the 2-D inverse transform to obtain the reconstructed image.

The third step requires a 2-D transform, so it should be performed after obtaining all the projection data.

5.2.3 Phantom Reconstruction

To test the correctness of the reconstruction formulas and to study the influence of various parameters of the reconstruction results, some ground-truth images are needed. Such a ground-truth image in medical image reconstruction of a head section is often called a phantom, as it is not from a real head. Different **phantoms** have been designed. An image that is often used is the Shepp-Logan head model (Shepp and Logan, 1974). A modified image (115×115, 256 gray levels) is shown in Figure 5.9, and all its parameters are listed in Table 5.1.

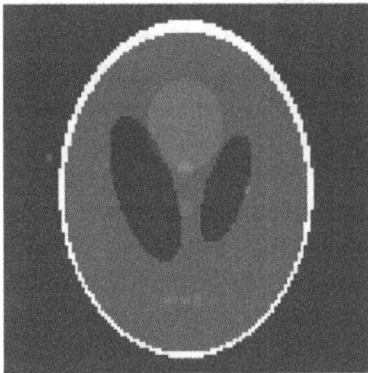

Figure 5.9: A modified image of a Shepp-Logan head model.

Table 5.1: Parameters of the modified Shepp-Logan head model.

Serial number of ellipses	X-axis coordinates of center	Y-axis coordinates of center	Half-length of short-axis	Half-length of long-axis	Angle between Y-axis and long-axis	Relative density
A (Big outside)	0.0000	0.0000	0.6900	0.9200	0.00	1.0000
B (Big inside)	0.0000	−0.0184	0.6624	0.8740	0.00	−0.9800
C (Right inclined)	0.2200	0.0000	0.1100	0.3100	−18.00	−0.2000
D (Left inclined)	−0.2200	0.0000	0.1600	0.4100	18.00	−0.2000
E (Top big)	0.0000	0.3500	0.2100	0.2500	0.00	0.1000
F (Middle-high)	0.0000	0.1000	0.0460	0.0460	0.00	0.1000
G (Middle-low)	0.0000	−0.1000	0.0460	0.0460	0.00	0.1000
H (Bottom-left)	−0.0800	−0.6050	0.0460	0.0230	0.00	0.1000
I (Bottom-middle)	0.0000	−0.6060	0.0230	0.0230	0.00	0.1000
J (Bottom-right)	0.0600	−0.6050	0.0230	0.0460	0.00	0.1000

Example 5.2 Reconstruction with the Fourier inverse transform

In real reconstruction, a large number of projections are often required. Figure 5.10 shows some projection examples. From Figures 5.10(a)–(f), 4, 8, 16, 32, 64, and 90 views or sets of projections are taken at equivalent angle steps, and the Fourier transform of each projection is calculated and plotted into a 2-D complex image. It can be seen that the energy distribution concentrates at the center with the increase in the projection numbers.

The corresponding reconstructed images are shown in Figures 5.11(a)–(f). The quality of the reconstructed image gradually improves as the number of projections increase. The image quality (consider both the inconsistency and the blurriness) for

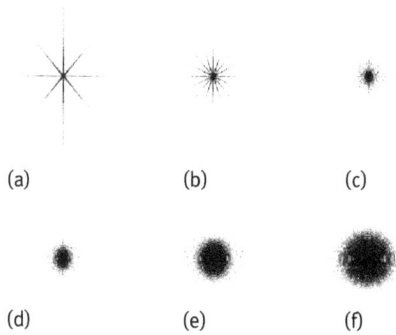

(a)   (b)   (c)

(d)   (e)   (f)

Figure 5.10: 2-D complex images obtained in reconstruction with the Fourier inverse transform.

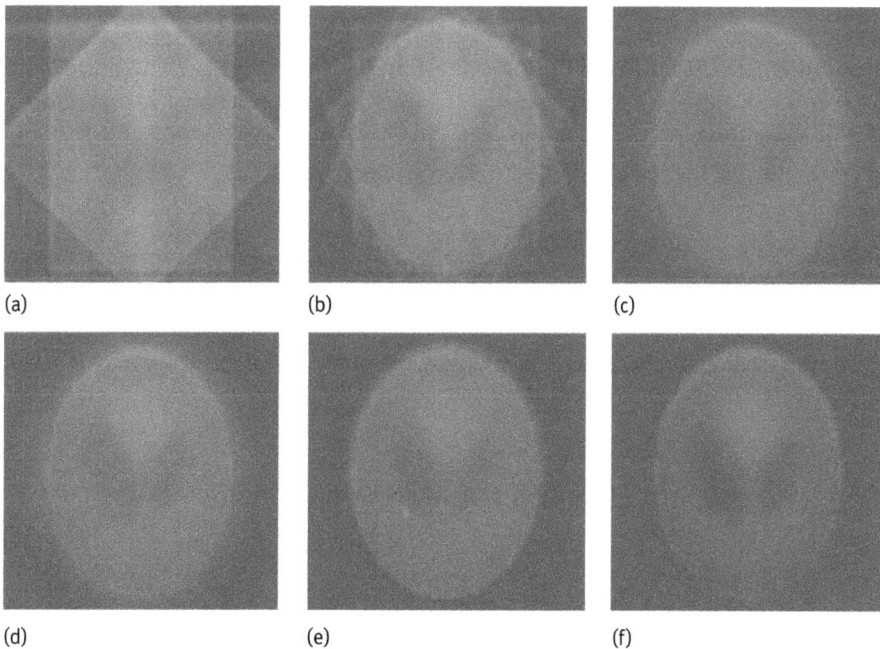

(a)   (b)   (c)

(d)   (e)   (f)

Figure 5.11: Reconstructed images with the Fourier inverse transform.

four views is quite poor in which even the biggest outer ellipse is not obvious. Owing to the limited number of the views, the image quality in Figures 5.11(b) and (c) is only fair, in which the small inner ellipses are not clear. When 90 views with an interval of two degrees are used, the result becomes quite good, as shown in Figure 5.11(f). However, because of the gaps in the frequency-space image, some artifacts are still present. In fact, all tomographic reconstruction procedures are sensitive to the number of views.

◻

5.3 Convolution and Back-Projection

Back-projection is used to project back the projections obtained from different directions to the original locations. If back projection is performed for each of these directions, it is possible to build a distribution on the plane. The most typical technique for back projection is convolution back-projection.

5.3.1 Convolution Back-Projection

Convolution back-projection can also be derived from the projection theorem for the Fourier transform. In the polar system, the inverse transform of eq. (5.21) is

$$f(x, y) = \int_0^\pi \int_{-\infty}^\infty G(R, \theta) \exp\left[j2\pi R(x \cos\theta + y \sin\theta)\right] |R| \, dR d\theta \qquad (5.29)$$

Substituting $G(R, \theta)$ by eq. (5.19) gives the formula for reconstructing $f(x, y)$ from $g(s, \theta)$. In a real application, as in reconstruction from the Fourier inverse transform, a window is required. According to the sampling theorem, the estimation of $G(R, \theta)$ can only be made in a limited band $|R| < 1/(2\Delta s)$. Suppose

$$h(s) = \int_{-1/(2\Delta s)}^{1/(2\Delta s)} |R| \, W(R) \exp[j2\pi Rs] dR \qquad (5.30)$$

Substituting it into eq. (5.29) and exchanging the order of the integration for s and R yields

$$f_W(x, y) = \int_0^\pi \int_{-1/(2\Delta s)}^{1/(2\Delta s)} G(R, \theta) W(R) \exp\left[j2\pi R(x \cos\theta + y \sin\theta)\right] |R| \, dR d\theta$$

$$= \int_0^\pi \int_{-1}^1 g(s, \theta) h(x \cos\theta + y \sin\theta - s) \, ds d\theta \qquad (5.31)$$

The above formula can be decomposed into two consecutive operations

$$g'(s, \theta) = \int_{-1}^{1} g(s, \theta)h(s' - s)ds \tag{5.32}$$

$$f_W(x, y) = \int_{0}^{\pi} g'(x \cos \theta + y \sin \theta, \theta)d\theta \tag{5.33}$$

where $g'(s', \theta)$ is the convolution of $h(s)$ and the projection of $f(x, y)$ along the θ direction, which is also called the convoluted projection along the θ direction with $h(s)$ as the convolution function. The process represented by eq. (5.32) is a convolution process, while the process represented by eq. (5.33) is a back-projection process. Since the parameters of $g'(\bullet)$ are the parameters of the line passing through (x, y) with an angle θ, $f_W(x, y)$ is the integration of the projections after convolution with all lines passing through (x, y).

5.3.1.1 Discrete Computation

The back-projection represented by eq. (5.33) can be approximated by

$$f_W(k\Delta x, l\Delta y) \approx \Delta\theta \sum_{n=1}^{N} g'(k\Delta x \cos \theta_n + l\Delta y \sin \theta_n, \theta_n) \tag{5.34}$$

For every θ_n, the values of $g'(s', \theta_n)$ for $K \times L$ values of s' are required. Direct computation is time consuming, as both K and L are quite large. One solution is to evaluate $g'(m\Delta s, \theta_n)$ for $M^- \leq m \leq M^+$, and then use interpolation to estimate the required $K \times L$ values of $g'(m\Delta s, \theta_n)$ from the M calculated values. In this way, the convolution in eq. (5.32) is approximated by two operations on discrete data. The first operation is a discrete convolution, whose result is denoted g'_C. The second operation is an interpolation operation, whose result is denoted g'_I. The following pair of equations shows the two operations

$$g'_C(m'\Delta s, \theta_n) \approx \Delta s \sum_{m=M^-}^{M^+} g(m\Delta s, \theta_n)h[(m' - m)\Delta s] \tag{5.35}$$

$$g'_I(s', \theta_n) \approx \Delta s \sum_{n=1}^{N} g'_C(m\Delta s, \theta_n)I(s' - m\Delta s) \tag{5.36}$$

where $I(\cdot)$ is an interpolating function.

Example 5.3 Illustration of convolution back-projection
A convolution can be considered a filtering process. In convolution back-projection, the data are first filtered and then the results are back projected. In this way, the blurriness is corrected. Figure 5.12 shows a set of examples using the convolution

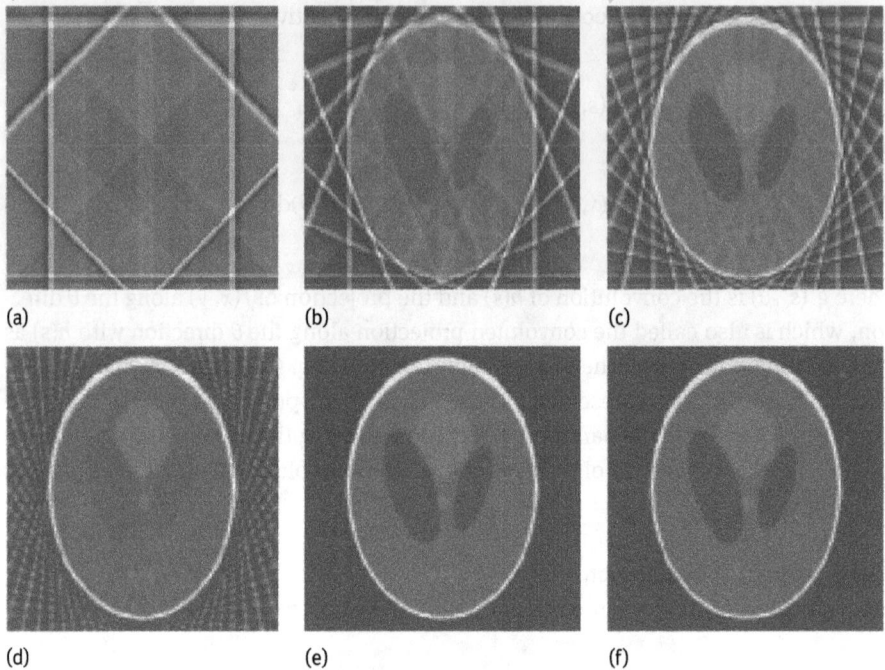

Figure 5.12: Reconstructed images with convolution back-projection.

back-projection with the model image of Figure 5.9. From Figure. 5.12(a)–(f), the reconstructed results are obtained with 4, 8, 16, 32, 64, and 90 views. In cases of only limited views, some bright lines along the projection directions are clearly visible, which are the results of the back-projection.

5.3.1.2 Reconstruction from Fan-Beam Projections

In real applications, it is necessary to collect the projection data in as short a time as possible in order to minimize distortions due to movement of the object as well as to avoid injuring patients. This is done by using fan-beam rays. They diverge from a ray source, pass through the object, and are received by an array of detectors. This mode of projection is called **fan-beam projection**. Two geometries can be distinguished, as depicted in Figure 5.13(a) and (b). In Figure 5.13(a), both the source and the detectors are attached in a frame and they move together in a circular path around the object (third-generation geometry). In Figure 5.13(b), the source moves, while the detectors are fixed in a ring form (fourth-generation geometry).

Reconstruction from the fan-beam projections can be converted to reconstruction from parallel rays by converting center projections to parallel projections.

Consider the geometry shown in Figure 5.13(c), in which the detectors are distributed evenly on an arc. Each ray (s, θ) is now one set of diverging rays (α, β), where α is the angle of divergence of the ray from the source-to-center line and β determines the

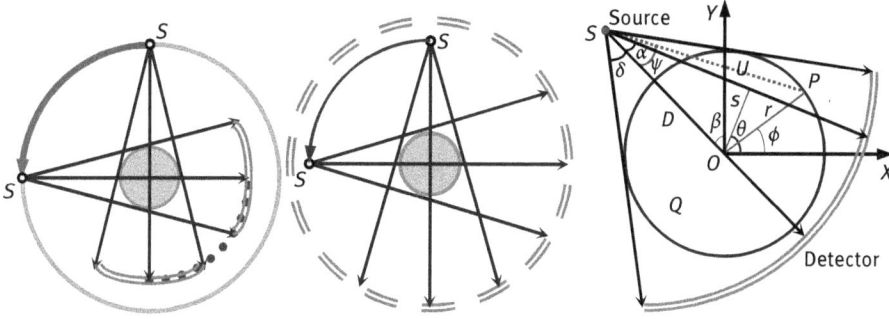

Figure 5.13: Fan-beam projection.

source position. The line integral $g(s, \theta)$ is denoted $p(\alpha, \beta)$ for $|s| < D$, where the locus of source positions is a circle of a radius D. It is assumed that the source is always located outside the object, so that $p(\alpha, \beta) = 0$ for $|\alpha| > \delta$ (δ is an acute angle).

From Figure 5.13(c), the following relations can be obtained

$$s = D \sin \alpha \quad \theta = \alpha + \beta \quad g(s, \theta) = p(\alpha, \beta) \tag{5.37}$$

In addition, suppose U is the distance from S to P (P is the point to be reconstructed, whose position can be specified by r and ϕ) and ψ is the angle between the line segments SO and SP, then

$$U^2 = [r\cos(\beta - \phi)]^2 + [D + r\sin(\beta - \phi)]^2 \tag{5.38}$$

$$\psi = \arctan \left[\frac{r\cos(\beta - \phi)}{D + r\sin(\beta - \phi)} \right] \tag{5.39}$$

Therefore,

$$r\cos(\theta - \phi) - s = U\sin(\psi - a). \tag{5.40}$$

According to eq. (5.37) to eq. (5.40), eq. (5.31) can be written as

$$f_W(r\cos\phi, r\sin\phi) = \frac{1}{2} \int\limits_{-\infty}^{\infty} \int\limits_{-\infty}^{\infty} \int\limits_{0}^{2\pi} g(s, \theta) \exp\{j2\pi R[r\cos(\theta - \phi - s)]\} W(R)|R| d\theta ds dR \tag{5.41}$$

Changing (s, θ) to (α, β), we get

$$f_W(r\cos\varphi, r\sin\varphi) = \frac{D}{2} \int\limits_{-\infty}^{\infty} \int\limits_{-\delta}^{\delta} \int\limits_{-a}^{2\pi-a} p(\alpha, \beta) \cos\alpha \exp[j2\pi RU \sin(\psi - \alpha)] W(R) |R| d\beta d\alpha dR \tag{5.42}$$

Note that the integrand is periodic of a period 2π in β, so that the limits of the β-integration may be replaced by 0 and 2π. The order of the integrations may be exchanged directly and the reconstruction formula in eq. (5.42) can be decomposed into two steps, as was done with the parallel-ray geometry (eq. (5.32) and eq. (5.33)). The two steps are:

$$p(\psi, \beta) = \int_{-\delta}^{\delta} p(\alpha, \beta)\, h[U \sin(\psi - \alpha)] \cos \alpha \, d\alpha \tag{5.43}$$

$$f_W(r \cos \phi, r \sin \phi) = \frac{D}{2} \int_0^{2\pi} p(\psi, \beta) d\beta \tag{5.44}$$

5.3.1.3 Comparison with Fourier Inversion

Both Fourier inversion and convolution back-projection are based on the projection theorem for the Fourier transform. One difference is that to derive the Fourier reconstruction algorithm the 2-D inverse Fourier transform is expressed in the Cartesian coordinates, while to derive the convolution back-projection algorithm, the 2-D inverse Fourier transform is expressed in the polar coordinates.

In view of their common history, it is surprising that these algorithms are utilized very differently in practice: The convolution back-projection algorithm and its divergent-ray offspring are almost universally adopted in CT, whereas the Fourier reconstruction algorithm is seldom used. One reason is that the basic algorithm of convolution back-projection is straightforward and implemented in software or hardware, and it produces sharp and accurate images from good-quality data. On the other hand, the Fourier method is difficult to implement, and the images it produces are remarkably inferior to those obtained by convolution back-projection, due to the inelegant 2-D interpolation required. Another reason is that the formula for the convolution back-projection derived from the parallel-ray case can be modified to suit the applications with fan-beam projections, as shown by eq. (5.42), whereas the formula for the Fourier inversion derived from the parallel-ray example does not keep the original efficiency when it is modified for applications with fan-beam projections.

However, one characteristic that the Fourier method has in its favor is that it requires potentially less computation to reconstruct the image, which will become increasingly important if the amount of data and the size of image matrices continues to increase in practical applications.

5.3.2 Filter of the Back-Projections

In the approach that uses a **filter of the back-projection**, the back-projection is applied first, followed by a filtering or a convolution (Deans, 2000). The back-projection produces a blurred image, which is the 2-D convolution between the real image and $1/(x^2 + y^2)^{1/2}$.

Let this blurred image be designated by

$$b(x, y) = \mathscr{B}[R_f(p, \theta)] = \int_0^\pi R_f(x \cos \theta + y \sin \theta, \theta)d\theta \tag{5.45}$$

where $\mathscr{B}$ denotes the back-projection.

The true image is related to $b(x, y)$ by (using two convolution symbols to represent 2-D convolution)

$$b(x, y) = f(x, y) \otimes \otimes \frac{1}{(x^2 + y^2)^{1/2}} = \int_{-\infty}^\infty \int_{-\infty}^\infty \frac{f(x', y')dx'dy'}{\left[(x - x')^2 + (y - y')^2\right]^{1/2}} \tag{5.46}$$

To see the relation more clearly, the following relation can be used

$$\mathscr{R}[f(x, y)] = \mathscr{F}_{(1)}^{-1}\mathscr{F}_{(2)}[f(x, y)] \tag{5.47}$$

where $\mathscr{R}$ denotes the Radon transform, $\mathscr{F}$ denotes the Fourier transform, and the subscript number in the parentheses denotes the dimension of the transform.

Applying the back-projection operator to both sides obtains

$$b(x, y) = \mathscr{B}\left[R_f(x, y)\right] = \mathscr{B}\mathscr{F}_{(1)}^{-1}\mathscr{F}_{(2)}[f(x, y)] \tag{5.48}$$

where the subscripts represent the dimension of the Fourier transform. Following the above order of operations, this method is better called back-projection of the filtered projections.

The inverse 1-D Fourier transform in eq. (5.48) is operated on a radial variable, which means $F(u, v)$ must be converted to the polar form $F(q, \theta)$ before performing the inverse 1-D Fourier transform. The variable q is the radial variable in the Fourier space, where $q^2 = u^2 + v^2$. If the inverse 1-D Fourier transform of $F(q, \theta)$ is designated by $f(s, \theta)$, then

$$b(x, y) = \mathscr{B}[f(s, \theta)] = \mathscr{B}\int_{-\infty}^\infty F(q, \theta) \exp(j2\pi sq)dq \tag{5.49}$$

Mapping s to $x \cos \theta + y \sin \theta$ yields

$$b(x, y) = \int_0^\pi \int_{-\infty}^\infty F(q, \theta) \exp[j2\pi q(x \cos \theta + y \sin \theta)]dqd\theta$$
$$= \int_0^{2\pi} \int_{-\infty}^\infty \frac{1}{q}F(q, \theta) \exp[j2\pi qz \cos(\theta - \phi)]qdqd\theta \tag{5.50}$$

Note that the expression on the right side of eq. (5.50) is just the inverse 2-D Fourier transform,

$$b(x, y) = \mathscr{F}_{(2)}^{-1} \left\{ |q|^{-1} F \right\} \tag{5.51}$$

From the convolution theorem, it can be derived that

$$b(x, y) = \mathscr{F}_{(2)}^{-1} \left\{ |q|^{-1} \right\} \otimes \otimes \mathscr{F}_{(2)}^{-1} \{F\} \tag{5.52}$$

The first term on the right side of eq. (5.52) gives $1/(x^2 + y^2)^{1/2}$, so eq. (5.46) is verified.

According to eq. (5.47), the reconstruction algorithm can be obtained by taking the 2-D Fourier transform, which is given by

$$\mathscr{F}_{(2)} \left[b(x, y) \right] = |q|^{-1} F(u, v) \tag{5.53}$$

or

$$F(u, v) = |q| \mathscr{F}_{(2)} \left[b(x, y) \right] \tag{5.54}$$

Substituting $b(x, y)$ with $\mathscr{B}[F(u, v)]$ and taking the inverse 2-D Fourier transform, the reconstruction formula for the filter of the back-projection is written by

$$f(x, y) = \mathscr{F}_{(2)}^{-1} \left\{ |q| \, \mathscr{F}_{(2)} \mathscr{B}[F(u, v)] \right\} \tag{5.55}$$

By introducing a 2-D window

$$G(u, v) = |q| \, W(u, v) \tag{5.56}$$

Equation (5.55) can be written as

$$\begin{aligned} f(x, y) &= \mathscr{F}_{(2)}^{-1} \left\{ G(u, v) \mathscr{F}_{(2)} \mathscr{B}[F(u, v)] \right\} \\ &= \mathscr{F}_{(2)}^{-1} \left[G(u, v) \right] \otimes \otimes \mathscr{B}[F(u, v)] = g(x, y) \otimes \otimes b(x, y) \end{aligned} \tag{5.57}$$

Once the window function is selected, $g(x, y)$ can be found in advance by calculating the inverse 2-D Fourier transform, and the reconstruction is accomplished by a 2-D convolution between the inverse 2-D Fourier transform and the back-projection of the projections. The block diagram for the implementation of the above algorithm is shown in Figure 5.14, where $\mathscr{R}$ denotes a Radon transform, $\mathscr{F}$ denotes the Fourier transform, and $\mathscr{B}$ denotes the back-projection.

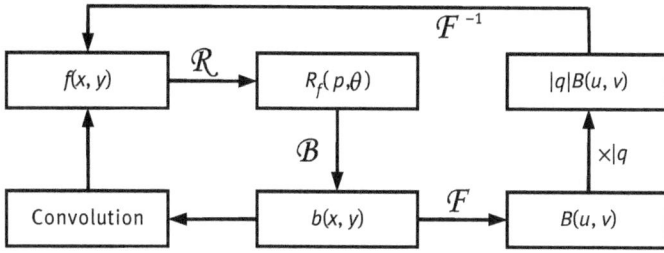

Figure 5.14: The diagram for the filter of the back-projection.

Example 5.4 Illustration of the filter of the back-projection.
A set of reconstructed images using the algorithm of the filter of the back-projection is illustrated in Figure 5.15. The original image is still the image shown in Figure 5.9. The images shown in Figures 5.15(a)–(f) correspond to the reconstructed results with 4, 8, 16, 32, 64, and 90 views, respectively. The quality of these images is better than those shown in Figure 5.11.

5.3.3 Back-Projection of the Filtered Projections

The algorithm known as the **filtered back-projection** algorithm is currently the optimum computational method for reconstructing a function from the knowledge of

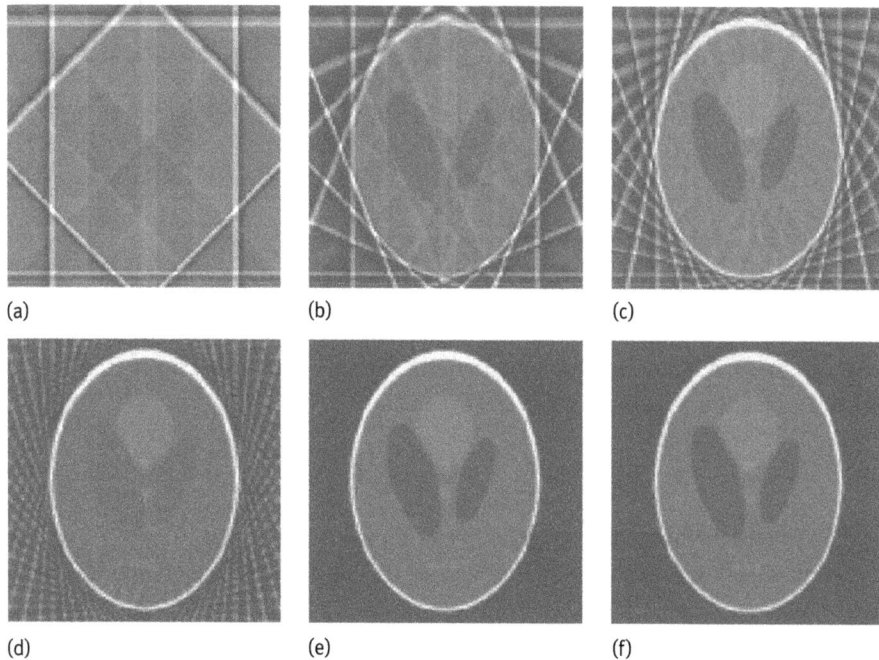

Figure 5.15: Reconstructed images with the filter of the back-projection.

its projections (Deans, 2000). The idea behind this algorithm is that the attenuation in each projection is dependent on the structure of the object along the projection direction. From a single projection, it is not possible to determine the attenuations on different points along the direction. However, the attenuation along a projection direction can be evenly distributed to all points along the direction. By performing this distribution for many directions and adding the attenuations from many directions, the attenuation for each point can be determined. This method is equivalent to the method of reconstruction in the Fourier space with many projections, but its computational cost is much less.

This algorithm can also be considered an approximate method for computer implementation of the inversion formula for the Radon transform. Since the filtering of the projections is performed before the back-projection, a better name is back-projection of the filtered projections. There are several ways to derive the basic formula for this algorithm. To emphasize its relation to the inversion formula, the starting point would be the back-projection.

Let $G(p, \theta)$ be an arbitrary function of a radial variable p and an angle θ. To obtain a function of x and y, the back-projection is defined by replacing p by $x \cos \theta + y \sin \theta$ and integrating over the angle θ,

$$g(x, y) = \mathscr{B}\left[G(p, \theta)\right] = \int_0^\pi G(x \cos \theta + y \sin \theta,\ \theta)d\theta \tag{5.58}$$

From the definition of back-projection, the reconstruction formula can be written as (Deans, 2000)

$$f(x, y) = \mathscr{B}\left[F(s, \theta)\right] = \int_0^\pi F(x \cos \theta + y \sin \theta,\ \theta)d\theta \tag{5.59}$$

where

$$F(s, \theta) = \mathscr{F}^{-1}\left\{|q|\,\mathscr{F}\left[R_f(p, \theta)\right]\right\} \tag{5.60}$$

The inverse Fourier transform converts a function of p to a function of some other radial variable s. A frequency implementation adds a window to approximate F before the back-projection, therefore, eq. (5.60) becomes

$$F(s, \theta) = \mathscr{F}^{-1}\left\{|q|\,W(q)\mathscr{F}\left[R_f(p, \theta)\right]\right\} \tag{5.61}$$

The diagram for the filtered back-projection is shown in Figure 5.16, where $\mathscr{R}$ denotes the Radon transform, $\mathscr{F}$ denotes the Fourier transform, and $\mathscr{B}$ denotes the back-projection.

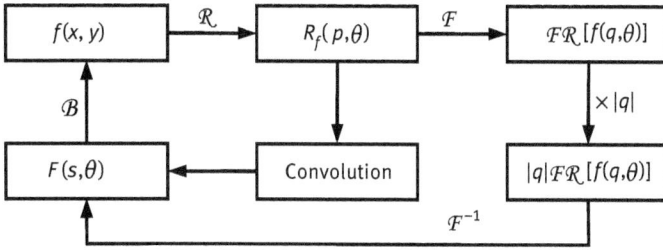

Figure 5.16: The diagram for the filter back-projection.

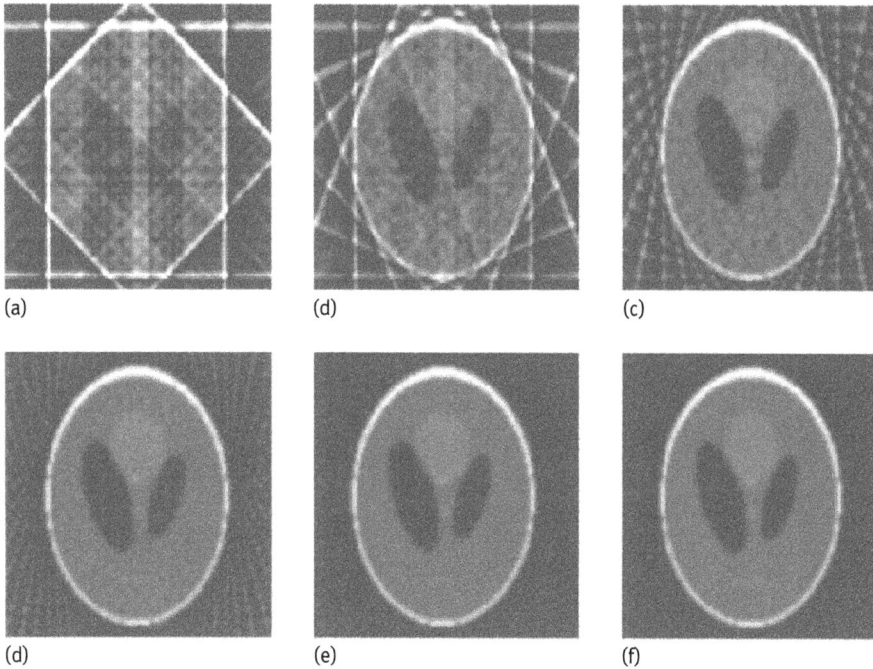

Figure 5.17: Reconstructed images with the filtered back-projection.

Example 5.5 Illustration of filtered back-projection
A set of reconstructed images using the algorithm of the filtered back-projection is illustrated in Figure 5.17. The original image is still the image shown in Figure 5.9. The images shown in Figures 5.17(a)–(f) correspond to the reconstructed results with 4, 8, 16, 32, 64, and 90 views, respectively. The quality of these images is similar to those shown in Figure 5.12, but have a higher contrast, especially for the cases with fewer views.

Note that with the increase in the number of views, the main structures of the reconstructed image becomes clearer and clearer, but the originally even regions become uneven (decrease from the center to the entourage). The main reason is that the projections from different directions overlap more in the center regions than the regions around.

5.4 Algebraic Reconstruction

Reconstruction is used to determine the density (**attenuation coefficient**) values of an object at every location. The input of a reconstruction system is the integration of the projections along each view. The summation of the attenuation coefficients from different projection lines at each pixel gives the measured attenuation values. As shown in Figure 5.18, the total attenuation value along the horizontal line is the summation of the attenuations at pixels $a_{11}, a_{21}, \ldots, a_{n1}$.

In Figure 5.18, each line integration provides an equation for a projection, and all these equations form a homogeneous system of equations. In this system of equations, the number of unknown variables is the number of pixels in the image and the number of the equations is the number of the line integrations that equals the product of the number of detectors along one direction of the projection and the number of the projections. Such a system has a large number of equations, but it is generally a sparse system as many pixels are not included in a particular line integration. One typical method used to solve this problem is called the **algebraic reconstruction technique** (ART) (Gordon, 1974) or the **finite series-expansion reconstruction method** (Censor, 1983). These are also called iterative algorithms or optimization techniques.

5.4.1 Reconstruction Model

Different from the previously described transform or projection methods, ART is performed in the discrete space from the beginning. Its principle can be explained with the illustration in Figure 5.19.

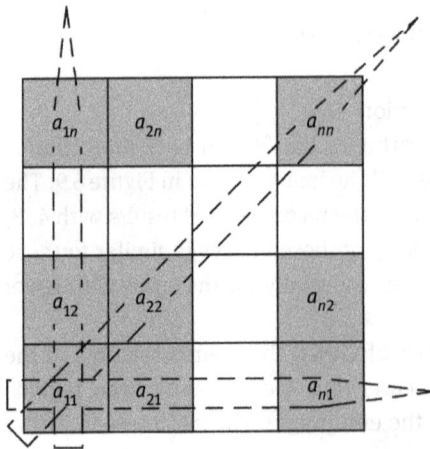

Figure 5.18: Line integrations cover a number of pixels.

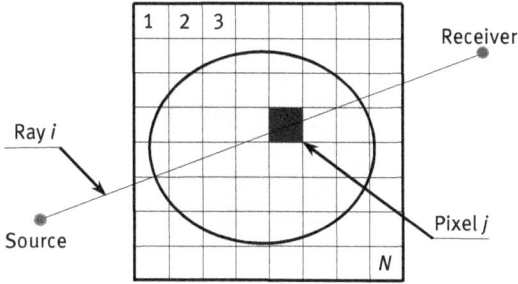

Figure 5.19: Illustration of ART.

In Figure 5.19, the object to be reconstructed is in a Cartesian grid. Both the source and the detectors are assumed to be point-like. The connecting line from the source to the detector corresponds to a radial line (total M lines). The pixels of the image are numbered from 1 to N (N denotes the total number of grids). The attenuation function is assumed to take a constant value x_j throughout the j-th pixel. The length of the intersection of the i-th ray with the j-th pixel is denoted a_{ij}. It represents the weight of the contribution of the j-th pixel to the total attenuation along the i-th ray.

If the measurement of the total attenuation of the i-th ray is denoted

$$y_i \approx \sum_{j=1}^{N} x_j a_{ij} \quad i = 1, 2, \cdots, M \tag{5.62}$$

Equation (5.62) can be written in a matrix form as

$$\mathbf{y} = \mathbf{A}\mathbf{x} \tag{5.63}$$

where $\mathbf{y}$ is the measurement vector, $\mathbf{x}$ is the image vector, and $\mathbf{A}$ is a $M \times N$ projection matrix.

5.4.2 Algebraic Reconstruction Technique

Theoretically, if the inverse matrix of the matrix $\mathbf{A}$ in eq. (5.63) could be obtained, then the image vector $\mathbf{x}$ could be calculated from the measurement vector $\mathbf{y}$. In practice, to obtain high-quality images, both M and N should be at least the order of 10^5. Therefore, $\mathbf{A}$ is normally a very large matrix. Taking an image of 256×256 as an example, to solve eq. (5.63), the numbers of rays should be equal to or bigger than the numbers of pixels in images, so $M \times N \approx 4.3 \times 10^9$. To put a matrix of such size totally inside the computer does already present a challenge. In addition, it has been proved that to compute the inverse of a matrix with D elements, the required computation time is proportional to $D^{3/2}$. For a 256×256 image, the required computation time will attend around 2.8×10^{14}. This is another challenge for the power of computer.

In practice, iterative techniques are often used to solve this problem. **Algebraic reconstruction technique** (ART) is a simple and typical iterative method. In its first step, an image vector $\mathbf{x}^{(0)}$ is initialized. The following iterative equation is then applied

$$x^{(k+1)} = x^{(k)} + \frac{y_i - a^i \cdot x^{(k)}}{||a^i||^2} a^i \tag{5.64}$$

where $a^i = (a_{ij})_{j=1}^n$ is a vector and "$\cdot$" denotes the inner product. In an iterative step, the current iterate $x^{(k)}$ is updated to $x^{(k+1)}$ by taking into account only a single ray and updating only the values of the pixels that intersect this ray. The difference between the measurement y_i and the pseudo-projection data $\sum_{j=1}^n a_{ij}x_j^{(k)}$ is redistributed among the pixels (along the i-th ray) proportionally to their weight a_{ij}.

ART uses an iterative procedure for each ray. It avoids the direct computation for inverse matrix, so the computation time is reduced. It also reduces the storage requirement as in each iteration only one line data in a matrix are utilized.

5.4.3 Simultaneous Algebraic Reconstruction Technique

Simultaneous algebraic reconstruction technique (SART) is an improvement of ART. ART uses the ray-by-ray fashion to iteratively update the reconstruction, that is, for each time a ray is computed, all pixel values related to this ray are updated. On the other side, SART makes the combined consideration of all rays related to a particular projection angle, that is, all measurements obtained from all rays passing through the same pixel are combined to update the pixel value (the results are independent to the order of ray measurements being used). In other words, a number of rays are combined in one iteration of SART, the utilization of the average value of these rays make it possible to achieve the result of suppression of interference factors.

The following iterative equation is applied:

$$x^{(k+1)} = x^{(k)} + \frac{\sum_{i\in I_\theta}\left[\frac{y_i - a^i \cdot x^{(k)}}{||a^i||^2}a^i\right]}{\sum_{i\in I_\theta} a^i} \tag{5.65}$$

where I_θ represents the set of rays corresponding to certain projection angles.

The main iterative steps of SART include:
(1) Initializing an image vector $x^{(0)}$ as the start point of iteration.
(2) Computing the projection value of i-th projection under a given projection angle θ.
(3) Computing the difference (residual) between the real measurement value and the projection value.
(4) $i = i + 1$, repeating step 2 to step 3, summarizing all projection differences under the same projection direction.
(5) Computing the modification value for image vector $x^{(k)}$.
(6) Revising image vector $x^{(k)}$ according to the modification value.
(7) $k = k + 1$, repeating step 2 to step 6, until passing through all projection angle, that is, complete one iteration.

(8)   Taking the previous iterative result as the initial value, repeating step 2 to step 7, until the algorithm converge.

It is seen from the above description that the number of sub-iterative processes in SART is much less than that in ART. Hence, SART is better for suppressing stripping artifact and for producing smoother reconstructed image.

Figure 5.20 shows a set of results obtained by the above method (Russ, 2002). Fan-beam geometry is used with 25 detectors evenly distributed around 1/4 of a circle.

(a)

5	5	5	5	5	5	5	5	5	5	5	5	5	5	5	5
5	5	5	5	5	5	5	5	5	5	5	5	5	5	5	5
5	5	5	5	5	5	5	5	5	5	5	5	5	5	5	5
0	0	0	0	0	20	20	20	20	20	20	0	0	0	0	0
0	0	0	0	0	20	20	20	20	20	20	0	0	0	0	0
0	0	0	0	0	20	20	20	20	20	20	0	0	0	0	0
0	0	0	0	0	20	20	20	20	20	20	0	0	0	0	0
0	0	0	0	0	20	20	20	20	20	20	0	0	0	0	0
0	0	0	0	0	20	20	20	20	20	20	0	0	0	0	0
5	5	5	5	5	5	5	5	5	5	5	5	5	5	5	5
5	5	5	5	5	5	5	5	5	5	5	5	5	5	5	5
5	5	5	5	5	5	5	5	5	5	5	5	5	5	5	5
5	5	5	5	5	5	5	5	5	5	5	5	5	5	5	5
5	5	5	5	5	5	5	5	5	5	5	5	5	5	5	5
5	5	5	5	5	5	5	5	5	5	5	5	5	5	5	5
5	5	5	5	5	5	5	5	5	5	5	5	5	5	5	5

(b)

13	19	14	3	0	0	6	8	8	4	0	0	0	9	9	9
0	12	12	3	2	0	7	7	7	6	0	0	0	3	12	12
0	1	5	5	3	3	11	10	9	8	1	2	2	1	8	11
0	0	0	0	3	10	14	14	14	14	6	2	2	1	2	10
0	0	0	0	4	16	19	16	15	15	8	1	2	1	2	6
0	2	4	3	5	13	16	17	17	15	10	5	2	3	2	2
5	3	2	2	3	13	15	15	16	15	14	4	6	3	2	3
2	2	2	2	3	10	11	12	15	16	16	10	4	5	5	4
2	2	2	2	4	10	11	12	13	15	16	11	8	7	3	3
2	0	0	0	0	2	5	6	8	9	11	8	4	5	6	7
3	0	0	0	4	4	5	6	7	9	10	11	5	4	3	3
3	1	0	3	5	4	4	6	7	7	9	9	6	5	7	4
3	3	1	1	5	4	5	5	7	8	8	9	5	4	4	6
5	4	3	4	7	6	5	5	5	7	8	9	5	5	5	6
4	5	5	6	6	6	6	5	6	6	7	8	9	5	4	4
4	5	6	8	6	6	6	5	5	5	8	8	7	5	3	4

(c)

6	13	9	0	2	4	9	8	8	4	0	0	0	8	10	10
0	7	9	4	3	1	9	6	5	4	0	0	1	2	12	13
0	3	4	2	2	2	13	12	9	8	1	0	0	1	6	11
4	0	0	0	3	11	16	17	16	15	6	0	0	0	0	8
0	0	0	1	5	19	23	16	14	14	9	1	1	0	1	3
0	0	3	4	5	15	17	18	17	14	11	4	1	2	2	1
4	0	2	2	4	13	17	15	16	15	14	3	4	2	1	4
2	1	2	1	3	12	11	13	15	15	16	10	3	3	4	3
1	3	0	1	4	11	11	12	14	14	16	11	8	5	2	1
3	2	1	0	1	2	5	5	6	7	8	7	2	2	5	6
4	1	1	0	5	4	5	6	8	8	10	12	5	4	2	1
5	2	1	3	5	3	4	6	7	7	7	10	6	4	5	4
3	4	2	2	5	4	5	6	7	7	7	9	5	4	3	4
6	3	2	5	8	5	5	4	5	6	7	9	5	4	5	5
5	7	5	6	6	5	6	5	6	5	6	6	9	5	3	3
5	7	8	10	6	5	6	3	4	4	7	7	7	4	3	3

(d)

3	10	5	1	4	4	9	9	8	5	1	1	2	7	8	8
1	5	10	5	3	0	8	5	4	3	0	1	1	2	10	10
1	2	3	1	1	2	11	12	9	7	0	0	1	3	5	8
1	0	0	0	2	10	18	17	17	15	6	1	0	0	0	5
0	0	0	1	6	21	24	17	15	15	10	1	0	0	0	3
0	0	3	3	6	16	20	17	16	16	12	4	2	2	1	1
2	1	2	0	3	12	19	15	16	17	13	4	3	2	1	3
0	1	1	2	2	13	11	14	15	15	17	9	4	3	4	3
1	3	0	0	3	12	11	13	14	14	17	11	8	5	3	2
2	1	0	0	1	0	3	6	7	8	8	9	2	2	5	6
2	1	2	0	5	3	5	6	8	9	10	12	6	4	2	1
4	2	1	3	5	3	4	7	8	6	7	10	6	4	5	5
2	4	2	2	5	4	5	6	8	7	7	9	5	4	4	4
7	3	1	5	8	5	6	5	6	5	7	9	5	4	5	5
7	8	6	6	7	5	7	4	6	5	5	6	10	5	3	4
7	8	11	12	7	6	6	1	3	4	8	7	7	5	3	3

Figure 5.20: Illustration of the iterative effects in ART.

The 16 × 16 array of the voxels has density values from 0 to 20 as shown in Figure 5.20(a). Three projection sets at view angles of 0°, 90°, and 180° are calculated for the reconstruction. In fact, this gives 75 equations with 256 unknown variables. The reconstruction is started from the initial density value 10. Figures 5.20(b)–(d) show the results obtained after 1, 5, and 50 iterations.

It can be seen from Figure 5.20 that the hollow regions and the internal squares are reconstructed quickly, while the boundaries are gradually enhanced. The errors are evident particularly at the corners of the image where a small number of rays are passing through, and at the corners of the internal square, where the attenuation value changes abruptly.

5.4.4 Some Characteristics of ART

Transform methods are faster than ART in reconstruction. Therefore, they have been used in most practical systems. However, ART has some useful properties (Censor, 1983).
(1)   New models for the reconstruction are formulated either with new underlying physical principles or by introducing new geometries of data collection. ART can be easily adapted to new geometries and new problems. ART is more flexible.
(2)   ART is more suitable for reconstruction of high-contrast objects (such as industrial objects).
(3)   ART performs generally better than transform techniques for reconstruction from a small number of directions (less than 10).
(4)   ART is more suitable than transform methods for ECT.
(5)   ART is more suitable than transform methods for 3-D reconstruction.
(6)   ART is more suitable than transform methods for incomplete reconstruction.

5.5 Combined Reconstruction

Combining the methods presented in the above three sections, some new techniques can be formulated. The combination can happen in the formula derivation, method implementation, or in applications. Some examples of combined reconstruction include:
(1)   Iterative transform.
(2)   Iterative reconstruction re-projection.
(3)   Reconstruction using angular harmonics.
(4)   Reconstruction by polynomial fit to projection.

In this section, the **iterative transform method** (also called continuous ART) is presented. The way in which the algorithm for discrete data is derived from the reconstruction formula qualifies it as a transform method, but its iterative nature and its representation of the image make the algorithm similar to the iterative algorithms derived by ART (Lewitt, 1983).

From Figure 5.5, suppose $f(x, y)$ is 0 outside of Q and $L(s, \theta_n)$ is the length of the intersect segment of line (s, θ_n) with Q. The task of reconstruction can be described as reconstructing a function $f(x, y)$, given its projections $g(s, \theta_n)$ for all real numbers s and a set of N discrete angles θ_n. The reconstruction formula operates on a "semi-discrete" function of $g(s, \theta_n)$ containing two variables, where θ_n is a discrete variable and s is a continuous variable.

For $i \geq 0$, the $(i + 1)$-th step of the algorithm produces an image $f^{(i+1)}(x, y)$ from the present estimation $f^{(i)}(x, y)$ given by

$$f^{(i+1)}(x, y) = \begin{cases} 0 & \text{if } f(x, y) \notin Q \\ f^{(i)}(x, y) + \frac{g(s, \theta_n) - g^{(i)}(s, \theta_n)}{L(s, \theta_n)} \end{cases} \tag{5.66}$$

where $n = (i \bmod N) + 1$, $s' = x \cos \theta_n + y \sin \theta_n$, $g^{(i)}(s, \theta_n)$ is the projection of $f^{(i)}(x, y)$ along the direction θ_n, and $f^{(0)}(x, y)$ is the given initial function. The sequence of images $f^{(i)}(x, y)$ generated by such a procedure is known to converge to an image which satisfies all the projections.

The physical interpretation of eq. (5.66) is as follows. For a particular angle θ_n, find the projection at angle θ_n of the current estimation of the image, and subtract this function from the given projection at angle θ_n to form the residual projection as a function of s for this angle. For each s of the residual, apply a uniform adjustment to all image points that lie on the line (s, θ_n), where the adjustment is such that $g^{(i+1)}(s, \theta_n)$ agrees with the given $g(s, \theta_n)$. Applying such an adjustment for all s results in a new image, which satisfies the given projection for this angle. The process is then repeated for the next angle in the sequence. It is often useful to view each step as a one-view back-projection of a scaled version of the residual.

Equation (5.66) specifies operations on "semi-discrete" functions. For a practical computation, an algorithm that operates on discrete data is required. Following the method used in Section 5.2, to estimate $g(s, \theta_n)$ from its samples $g(m\Delta s, \theta_n)$, an interpolating function $q(\cdot)$ containing one variable is introduced

$$g(s, \theta_n) \approx \sum_{m=M^-}^{M^+} g(m\Delta s, \theta_n) q(s - m\Delta s) \tag{5.67}$$

Similarly, to estimate $f(x, y)$ from its samples $f(k\Delta x, l\Delta y)$, a basis function $B(x, y)$ can be introduced that acts as an interpolating function of the two variables

$$f(x, y) \approx \sum_{k=K^-}^{K^+} \sum_{l=L^-}^{L^+} f(k\Delta x, l\Delta y) B(x - \Delta x, y - l\Delta y) \tag{5.68}$$

Replacing $f(x, y)$ by $f^{(i)}(x, y)$ in eq. (5.68) and substituting this expression in eq. (5.15), it gets

$$g^{(i)}(s, \theta) = \sum_{k,l} f^{(i)}(k\Delta x, l\Delta y) G_{k,l}^{(B)}(s, \theta) \qquad (5.69)$$

where

$$G_{k,l}^{(B)}(s, \theta) = \int B(s \times \cos\theta - t \times \sin\theta - k\Delta x, s \times \sin\theta + t \times \cos\theta - l\Delta y) dt \qquad (5.70)$$

According to the principles of the interpolation stated above, the continuous ART can be discretized. Deriving from eq. (5.67) and eq. (5.69), eq. (5.66) can be approximated as

$$f_{k,l}^{(i+1)} = \begin{cases} 0 & \text{if } (k\Delta x, l\Delta y) \notin Q \\ f_{k,l}^{(i)} + \dfrac{\sum_m \left[g(m\Delta s, \theta_n) - \sum_{k,l} f_{k,l}^{(i)} \times G_{k,l}^{(B)}(m\Delta s, \theta_n) \right] q[s_{k,l}(\theta_n) - m\Delta s]}{L[s_{k,l}(\theta_n), \theta_n]} \end{cases} \qquad (5.71)$$

where $n = (i \bmod N) + 1$, $s_{k,l}'(\theta) = (k\Delta x) \cos\theta + (l\Delta y) \sin\theta$, and

$$f_{k,l}^{(i)} = f^{(i)}(k\Delta x, l\Delta y) \qquad (5.72)$$

5.6 Problems and Questions

5-1 Given two 3×3 images as shown in Figure Problem 5-1,

(1) Compute five parallel projections along each of the three directions with $\theta = 0°, 45°$, and $90°$, respectively. The distance between two adjacent parallel lines is 1, and the centered projection passes through the center of the image.

(2) Can the image be reconstructed from these projections?

1	2	3		9	8	7
8	9	4		2	1	6
7	6	5		3	4	5

Figure Problem 5-1

5-2 A 5×5 matrix is composed of elements computed from $99/[1 + (5|x|)^{1/2} + (9|y|)^{1/2}]$, which are rounded to the nearest integer. The values of x and y are from -2 to 2. Compute the projections along the four directions with $\theta = 0°, 30°, 60°$, and $90°$.

5-3 Reconstructing a density distribution $f(x, y)$ from three projections is certainly not enough. Some people believe that if *a priori* knowledge $f(x, y) = \Sigma\delta(x - x', y - y')$ were available, the exact reconstruction from just three projections would be possible. However, this rather restrictive condition implies that all point masses are equal and none of them are negative. Others say, "If you can go that far, can't you handle positive impulse patterns of unequal strength?" What is your opinion?

5-4 The gray levels of a 4×4 image are shown in Figure Problem 5-4, four of which are known (the question mark means unknown). In addition, its projection on the X axis is $P(x) = [? ? 15 5]^T$, and on the Y axis it is $P(y) = [9 ? 27 ?]^T$. Can you reconstruct the original gray-level image?

?	3	?	?
0	?	?	?
6	?	?	?
?	3	?	?

Figure Problem 5-4

5-5 Printed material in 7×5 dot-matrix capitals, as shown in Figure Problem 5-5, is to be read automatically by scanning with a slit that yields the vertical projection of the letter. However, it is noted that several letter pairs have the same projection; for example, M and W, as well as S and Z. It is thought that an additional horizontal projection would eliminate the ambiguity.
(1) Is it true that the two projections would be sufficient?
(2) Do you have any suggestions for perfecting the scheme?

Figure Problem 5.5

5-6 Prove the following statements:

(1) If $f(x, y)$ is circularly symmetric, then it can be reconstructed from a single projection.

(2) If $f(x, y)$ can be decomposed into the product of $g(x)$ and $h(y)$, then it can be reconstructed from two projections that are perpendicular to two coordinate axes, respectively.

5-7* Prove the validity of the projection theorem for Fourier transforms.

5-8 Use the projection data obtained in Problem 5-1 to reconstruct the original images with reconstruction from Fourier inversion.

5-9 What are the relations and differences among the methods of reconstruction based on back projection?

5-10* Suppose that the linear attenuation coefficient at point (x, y) is $d(x, y)$. According to the attenuation law for X-ray,

$$\int_{(s,\theta)} d(x, y) = \ln\left(\frac{I_r}{I}\right)$$

where I_r is the intensity of the incident ray, I is the intensity after the incident ray passing through an object, and (s, θ) is the cross region of the line from the source to the receiver. Define the CT value as

$$CT = k\frac{d(x, y) - d_w(x, y)}{d_w(x, y)}$$

where $d_w(x, y)$ is the linear attenuation coefficient of water and k is a normalization coefficient. When using the technique of convolution back-projection with the impulse response of the reconstruction filter denoted $h(s)$, give the expression for the computing of CT.

5-11 Implement the ART with relaxation technique in Section 5.4 and verify the data in Figure 5.20.

5-12 Perform a reconstruction according to the projection data from Problem 5-1 using ART without relaxation and ART with relaxation, respectively. Compare the results with the original images in Problem 5-1.

5.7 Further Reading

1. Modes and Principles

- In history, the first clinical machine for detection of head tumors, based on CT, was installed in 1971 at the Atkinson Morley's Hospital, Wimbledon, Great Britain (Bertero and Boccacci, 1998). The announcement of this machine, by G.H. Hounsfield at the 1972 British Institute of Radiology annual conference, is considered the greatest achievement in radiology since the discovery of X-rays in 1895.

- In 1979, G.H. Hounsfield shared with A. Cormack the Nobel Prize for Physiology and Medicine (Bertero and Boccacci, 1998). The Nobel Prizes for Chemistry of 1981 and 1991 are also related to it (Herman, 1983), (Bertero and Boccacci, 1998), and (Committee, 1996).

- The phenomenon of nuclear resonance was discovered in 1946 by F. Bloch and M. Purcell, respectively. They shared the Nobel Prize for Physics in 1952 (Committee, 1996). The first MRI image was obtained in 1973.

- Complementary reading on the history and development of image reconstruction from projections can be found in Kak and Slaney (2001).

2. **Reconstruction by Fourier Inversion**

- An introduction and some examples of reconstruction by Fourier inversion can be found in Russ (2002).

- Another reconstruction method using 2-D Fourier transform is rho-filtered layergrams Herman (1980).

- More discussion and applications of phantoms can be found in Moretti *et al.* (2000).

- Another head phantom for testing CT reconstruction algorithms can be found in Herman (1980).

3. **Convolution and Back-Projection**

- The advantages of convolution and back-projection include that they are more suitable than other methods for data acquisition using a parallel mode (Herman, 1980).

- A recent discussion on back-projection algorithms and filtered back-projection algorithms can be found in Wei *et al.* (2005).

- More mathematical expressions on **cone-beam geometry** and 3-D image reconstruction can be found in Zeng (2009).

4. **Algebraic Reconstruction**

- The methods for algebraic reconstruction have many variations and further details can be found in Herman (1980), Censor (1983), Lewitt (1983), Zeng (2009), and Russ (2016).

5. **Combined Reconstruction**

- More combined reconstruction techniques, particularly second-order optimization for reconstruction and the non-iterative finite series-expansion reconstruction methods can be found in Herman (1980, 1983).

6 Image Coding

Image coding comprises a group of image processing techniques, and is also called image compression. Image coding addresses the problem of reducing the amount of data required to represent an image.

The sections of this chapter are arranged as follows:

Section 6.1 introduces some basic concepts and terms concerning image coding, such as data redundancy, image quality, and judgment.

Section 6.2 describes several commonly used variable length coding methods to reduce coding redundancy, including Huffman coding and its simplified versions, Shannon-Farnes coding, and Arithmetic coding.

Section 6.3 focuses on the methods of bit plane coding. The methods of decomposing the image, including binary decomposition and grayscale decomposition, are presented. Then the basic coding methods of binary image, including constant block coding, 1-D run-length coding, and 2-D run-length coding, are discussed.

Section 6.4 describes the currently used predictive coding methods, including lossless predictive coding and lossy predictive coding.

Section 6.5 presents the transform coding technique, which is a frequency domain and is lossy in general. Both the orthogonal transform coding and system and the wavelet transform coding and system are discussed.

6.1 Fundamentals

Some fundamental concepts and terms concerning image coding, such as data redundancy, image quality, and judgment, are first discussed.

6.1.1 Data Redundancy

Data are used to represent information. Various amounts of data may be used to represent the same amount of information. If one method uses more data than the others, then some data must provide nonessential information or already known information. In this case, there will be **data redundancy**.

Data redundancy can be quantitatively described. If n_1 and n_2 denote the number of information-carrying units in the two data sets that represent the same information, the **relative data redundancy** of the first data set can be defined as

$$R_D = 1 - \frac{1}{C_R} \tag{6.1}$$

DOI 10.1515/9783110524116-006

Table 6.1: Some particular values of C_R and R_D.

n_1 vs. n_2	CR	RD	Corresponding case
$n_1 = n_2$	1	0	The first data set contains no redundant data relative to the second data set.
$n_1 \gg n_2$	$\to \infty$	$\to 1$	The first data set contains much redundant data.
$n_1 \ll n_2$	$\to 0$	$\to -\infty$	The second data set contains much more data than the original data.

where C_R is called the **compression ratio** and is defined by

$$C_R = \frac{n_1}{n_2} \tag{6.2}$$

In the above two equations, C_R and R_D take values in open intervals of $(0, \infty)$ and $(-\infty, 1)$, respectively. Some particular values of C_R and R_D are shown in Table 6.1.

There are three basic kinds of data redundancy: coding redundancy, inter-pixel redundancy, and psycho-visual redundancy. Data compression is achieved when one or more of these redundancies are reduced or eliminated.

6.1.1 Coding Redundancy
For image coding, a code is established to represent the image data. A **code** is a system of symbols (letters, numbers, etc.) used to represent certain information. Each piece of information is assigned a sequence of code symbols, called a **code word**. The number of symbols in each code word is the length of the code word.

Suppose that a random variable s_k in the interval $[0, 1]$ represents the (normalized) gray levels of an image. Each s_k occurs with a probability $p_s(s_k)$

$$p_s(s_k) = n_k/n \quad k = 0, 1, \cdots, L-1 \tag{6.3}$$

where L is the number of gray levels, n_k is the total number of times that the k-th gray level appears in the image, and n is the total number of the pixels in the image. If the number of bits used to represent each value of s_k is $l(s_k)$, the average number of bits required to represent each pixel is

$$L_{avg} = \sum_{k=0}^{L-1} l(s_k)p_s(s_k) \tag{6.4}$$

According to eq. (6.4), if few bits are used to represent the gray levels with higher probability and more bits are used to represent the gray levels with lower probability, the **coding redundancy** can be reduced and the compression of the data can be achieved. If the gray levels of an image are coded in a way that uses more code symbols than are absolutely necessary to represent each gray level, the resulting image is said to contain coding redundancy.

6.1.1.2 Inter-Pixel Redundancy

Consider the images shown in Figures 6.1(a) and (b). Both have a series of identical circles, and they also have nearly the same histograms. However, the correlations among pixels in the two images are quite different. Here, the correlation can be estimated along a line of the image by

$$A(\Delta n) = \frac{1}{N - \Delta n} \sum_{x=0}^{N-1-\Delta n} f(x, y) f(x + \Delta n, y) \qquad (6.5)$$

where the variable y denotes the coordinate of the line and $\Delta n < N$. Figures 6.1(c) and (d) plot the functions of Δn along the center horizontal line of Figures 6.1(a) and (b), respectively.

Note that these plots have been normalized to facilitate the comparison by

$$A_0(\Delta n) = A(\Delta n)/A(0) \qquad (6.6)$$

The above example shows that there is another data redundancy, **inter-pixel redundancy**, which is directly related to the relationship between pixels in an image. Inter-pixel redundancy is also called spatial redundancy or geometric redundancy. Since the circles in Figure 6.1(a) are regularly arranged, the gray levels of many pixels can be reasonably predicted from other pixels, so the information carried by individual pixels should be less than that carried by pixels in Figure 6.1(b). In other words, Figure 6.1(a) has more inter-pixel redundancy than Figure 6.1(b).

6.1.1.3 Psycho-Visual Redundancy

People have different perceptions of the same scene, and their eyes do not respond with equal sensitivity to all visual information. Certain information may have less importance than other information and may be considered psycho-visual redundant. This information may be removed without significant damage to the image quality.

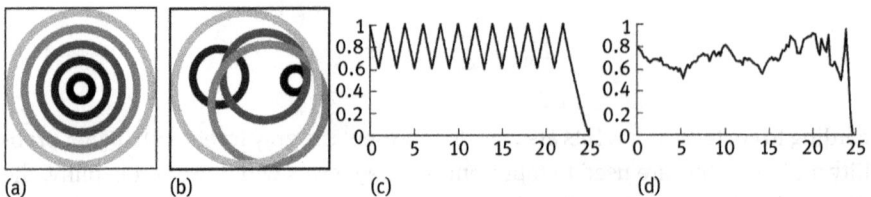

(a) (b) (c) (d)

Figure 6.1: Illustrations of original images and their auto-correlations.

Psycho-visual redundancy is different from the above two types of data redundancy. It is associated with visual information that is less significant to human perception but not redundant indeed, while coding redundancy and inter-pixel redundancy are the real types of redundancy. Elimination of the psycho-visual redundancy results in a loss of information, so it is only used for information that is not essential for normal visual processing.

6.1.2 Image Quality and Judgment

To achieve data compression, data redundancies should be reduced or removed. Removal of information will possibly produce the difference between the original image to be compressed and the compressed and decompressed image. If no difference is produced, the coding process is called a **lossless coding**. If some difference is produced, the coding is called a **lossy coding**. How to judge the fidelity of the decompressed image compared to the original image, more generally how to judge the quality of images, is an important problem. Two types of methods and two groups of criteria are discussed below (for other methods and criteria, see [Cosman et al., 1994]).

6.1.2.1 Objective Fidelity Criteria
It is often desirable that the information lost can be expressed as a function of the original image and the decompressed image. In this case, it is said that the expression is based on an **objective fidelity criterion**.

A simple criterion is the root-mean-square (rms) error between the image to be coded (input image) and the coded and then decoded image (output image). Let $f(x, y)$ represent an input image and $\hat{f}(x, y)$ represent an output image. For any value of x and y, the error $e(x, y)$ between $f(x, y)$ and $\hat{f}(x, y)$ is

$$e(x, y) = \hat{f}(x, y) - f(x, y) \tag{6.7}$$

Suppose that the sizes of the two images are both $M \times N$, then the total error between these two images is

$$\sum_{x=0}^{M-1} \sum_{y=0}^{N-1} \left| \hat{f}(x, y) - f(x, y) \right| \tag{6.8}$$

The rms error between these two images is

$$e_{\text{rms}} = \left[\frac{1}{MN} \sum_{x=0}^{M-1} \sum_{y=0}^{N-1} \left[\hat{f}(x, y) - f(x, y) \right]^2 \right]^{1/2} \tag{6.9}$$

Another objective fidelity criterion is the *mean-square signal-to-noise ratio* (SNR_{ms}). If $\hat{f}(x, y)$ is considered the sum of $f(x, y)$ and a noise image $e(x, y)$, then the mean-square SNR of the output image is given by

$$SNR_{ms} = \sum_{x=0}^{M-1} \sum_{y=0}^{N-1} \hat{f}(x, y)^2 \bigg/ \sum_{x=0}^{M-1} \sum_{y=0}^{N-1} \left[\hat{f}(x, y) - f(x, y)\right]^2 \qquad (6.10)$$

The root value of SNR_{ms}, denoted SNR_{rms}, is obtained by taking the square root of SNR_{ms}.

In a real application, SNR_{ms} is often normalized and represented by the unit decibel (dB) (Cosman *et al.*, 1994). Let

$$\bar{f} = \frac{1}{MN} \sum_{x=0}^{M-1} \sum_{y=0}^{N-1} f(x, y) \qquad (6.11)$$

then

$$SNR = 10 \log \left[\frac{\displaystyle\sum_{x=0}^{M-1} \sum_{y=0}^{N-1} \left[f(x, y) - \bar{f}\right]^2}{\displaystyle\sum_{x=0}^{M-1} \sum_{y=0}^{N-1} \left[\hat{f}(x, y) - f(x, y)\right]^2} \right] \qquad (6.12)$$

Let $f_{max} = \max \{f(x, y), x = 0, 1, \ldots, M-1, y = 0, 1, \ldots, N-1\}$, and the **peak SNR** (PSNR) is given by

$$PSNR = 10 \log \left[\frac{f_{max}^2}{\dfrac{1}{MN} \displaystyle\sum_{x=0}^{M-1} \sum_{y=0}^{N-1} \left[\hat{f}(x, y) - f(x, y)\right]^2} \right] \qquad (6.13)$$

6.1.2.2 Subjective Fidelity Criteria

Since most decompressed images ultimately are viewed by human eyes, measuring the image quality by the subjective evaluations of a human observer is also useful. One commonly used procedure is to show some typical images to a group of human evaluators (normally more than 20 persons are selected) and average their evaluation scores.

The evaluation is made with absolute rating scales. Table 6.2 shows one possible scale for television images.

The evaluation may also be made with some relative rating scales with side-by-side comparison. An example scale is $\{-3, -2, -1, 0, 1, 2, 3\}$, which can be used to represent the subjective assessments of {much worse, worse, slightly worse, the same, slightly better, better, much better}, respectively.

Table 6.2: Ratings of the image quality for television.

Value	Rating	Description
1	Excellent	An image of extremely high quality, as good as you could desire.
2	Fine	An image of high quality, providing enjoyable viewing. Interference is not objectionable.
3	Passable	An image of acceptable quality. Interference is not objectionable.
4	Marginal	An image of poor quality; you wish you could improve it. Interference is somewhat objectionable.
5	Inferior	A very poor image, but you could watch it. Objectionable interference is definitely present.
6	Unusable	An image so bad that you could not watch it.

The use of **subjective fidelity criteria** is often more complicated than objective fidelity criteria. In addition, the results obtained with subjective fidelity criteria do not always correspond to those obtained with objective fidelity criteria.

6.2 Variable-Length Coding

Variable-length coding is also called **entropy coding**, which is based on statistical theories. It can reduce the coding redundancy and keep the original information (lossless or error-free). The principle is to represent gray levels with high probability with fewer bits and to represent gray levels with low probability with more bits (variable-length), in order to get the minimum of eq. (6.4).

6.2.1 Huffman Coding

Huffman coding is a popular technique for reducing coding redundancy. Given a source $A = \{a_1, a_2, a_3, a_4\}$ with probability $\boldsymbol{u} = [0.1, 0.2, 0.3, 0.4]$, a Huffman code can be constructed by iteratively constructing a binary tree as follows:

(1) Arrange the symbols of A such that the probabilities of a_j are in ascending order, and consider the arranged symbols as the leaf nodes of a tree, as shown in Figure 6.2(a).

(2) Take the two nodes with the least two probabilities and merge them into a new node whose probability is the sum of the two probabilities, as shown in Figure 6.2(a). The new node is viewed as the parent node of the two nodes and it is connected to its child nodes. Assign 1 to one of the connection and 0 to the other connection randomly.

(3) Update the tree by replacing the two least probable nodes with their parent node, and reorder the nodes if needed. Repeat step 2 if there is more than one node, as shown in Figure 6.2(b). The last node is the root node of the tree, as in Figure 6.2(c).

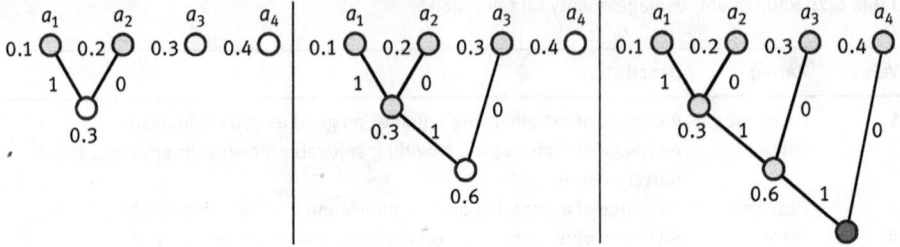

Figure 6.2: Huffman coding example.

(4) The code word of a symbol a_j can be obtained by traversing the linked path from the root node to the leaf nodes while reading sequentially the bit values assigned to the connections along the path. From Figure 6.2(c), the code words for **A** are {111, 110, 10, 0}.

6.2.2 Suboptimal Huffman Coding

According to the principle of Huffman coding, for a source with J symbols, the process of updating the coding tree should be performed with $J - 2$ times. To reduce the computation, it is often desired to sacrifice some coding efficiency for simplicity in code construction. Two common **suboptimal variable length coding** schemes are truncated Huffman coding and shift Huffman coding.

Consider that the information source is the histogram shown in Figure 5.13(a). Its symbols and probabilities are listed in the second and third columns of Table 6.3, respectively. For comparison, the Huffman code is listed in the fourth column.

The **truncated Huffman coding** is a simple modification of Huffman coding. It only codes the most probable M symbols with the Huffman code while codes the rest of the symbols by a fixed-length code with an additional prefix code. In Table 6.3, M is taken as 4. That is, the first four symbols are coded with the Huffman code, and

Table 6.3: Comparison of Huffman code and its modifications.

Block	Symbol	Probability	Huffman	Truncated Huffman		Shift Huffman	
Block 1	a_1	0.25	01	01	01	01	01
	a_2	0.21	10	10	10	10	10
	a_3	0.19	11	000	11	000	11
	a_4	0.16	001	001	001	001	001
Block 2	a_5	0.08	0001	11 00	000 00	11 01	000 01
	a_6	0.06	00000	11 01	000 01	11 10	000 10
	a_7	0.03	000010	11 10	000 10	11 00	000 11
	a_8	0.02	000011	11 11	000 11	11 001	000 001
	Average length		2.73	2.78		2.7	2.75

the other four symbols are merged to a special symbol, which is also coded with the Huffman code. Then the code for the special symbol is used as a prefix code for the four symbols forming the special symbol. In this example, the code for the special symbol can be either 11 or 000, depending on the code assigned to the symbol a_3.

The **shift Huffman coding** is also a simple modification of the Huffman coding. The process consists of the following steps:
(1)   Arrange the source symbols so that their probabilities are in descending order.
(2)   Divide the total number of symbols into blocks of equal size.
(3)   Code the elements within each block identically.
(4)   Add special shift-up and/or shift-down symbols to identify each block.

In Table 6.3, the total number of symbols is divided into two blocks. The first block is used as a reference block, and it is coded with all symbols in block 2 counted as a shift symbol. So the coding of the first block is similar to the truncated Huffman coding. The code for the shift symbol is then added to the symbols in block 2. Depending on the code assigned to the symbol a_3, two different shift codes are obtained.

By comparison, it is noted that the average lengths of both the truncated Huffman code and the shift Huffman code are (slightly) larger than the average length of the Huffman code. However, their computation is much faster than Huffman coding.

6.2.3 Shannon-Fano Coding

Shannon-Fano coding is also a variable-length coding technique. In Shannon-Fano code, 0 and 1 are independent and they appear with nearly the same probabilities. The coding process consists of the following steps:
(1)   Arrange the symbols of A such that the probabilities of a_j are in descending order.
(2)   Divide those source symbols, which have not been coded, into two parts, in which the sums of the probabilities for these two parts are as close as possible to each other.
(3)   Assign values 0 and 1 (or 1 and 0) to these two parts.
(4)   If both of these two parts have only one source symbol, stop the process. Otherwise, step (2) is repeated.

The above process can be illustrated with the following example. Suppose that a source $A = \{a_1, a_2, a_3, a_4, a_5, a_6\}$ with a probability $u = [0.1, 0.4, 0.06, 0.1, 0.04, 0.3]^T$ is used. The Shannon-Fano coding can give two results as shown in Tables 6.4 and 6.5, respectively. Though the coding results are different with different code words, their average lengths are the same.

Table 6.4: Shannon-Fano coding example 1.

Original source		Assignment for source symbols					Code word
Symbols	probability	1	2	3	4	5	
a_2	0.4	0					0
a_6	0.3		0				10
a_1	0.1			0			110
a_4	0.1	1			0		1110
a_3	0.06		1			0	11110
a_5	0.04			1	1	1	11111

Table 6.5: Shannon-Fano coding example 2.

Original source		Assignment for source symbols				Code word
Symbols	probability	1	2	3	4	
a_2	0.4	0				0
a_6	0.3		0			10
a_1	0.1			0	0	1110
a_4	0.1	1			1	1101
a_3	0.06		1		0	1110
a_5	0.04			1	1	1111

It has been proven that for a given source $A = \{a_1, \ldots, a_j, \ldots, a_J\}$ with a probability $u = [P(a_1), \ldots, P(a_j), \ldots, P(a_J)]^T$, if the length of the code word for a_j is L_j and the following two equations are satisfied

$$P(a_j) = 2^{-L_j} \tag{6.14}$$

$$\sum_{j=1}^{J} 2^{-L_j} = 1 \tag{6.15}$$

the efficiency of Shannon-Fano coding will attain its maximum, 100%. For example, for a given source $A = \{a_1, a_2, a_3\}$ with a probability $u = [1/2, 1/4, 1/4]^T$, the code words are $\{0, 10, 11\}$.

6.2.4 Arithmetic Coding

Arithmetic coding assigns an entire sequence of the source symbols to a single arithmetic code word. This code word defines an interval of the real numbers between 0 and 1. As the number of the symbols increases, the interval used to represent it becomes smaller and the number of information units required to represent the interval becomes larger.

Arithmetic coding generates non-block codes; that is, it does not require that each source symbol is translated into an integral number of code symbols. There is no one-to-one correspondence between the source symbols and the code words.

The coding process can be illustrated by an example. Again, take the source used in Figure 6.2 as an example; that is, a source $A = \{a_1, a_2, a_3, a_4\}$ with a probability $u = [0.1, 0.2, 0.3, 0.4]^T$. A five-symbol sequence $b_1 b_2 b_3 b_4 b_5 = a_4 a_3 a_2 a_1 a_3$ is coded as shown in Figure 6.3.

Suppose that the sequence occupies the entire half-open interval $[0, 1)$ at the beginning. This interval is subdivided into four regions based on the probability of each source symbol. That is, a_1 is associated with the subinterval $[0, 0.1)$, a_2 is associated with the subinterval $[0.1, 0.3)$, a_3 is associated with the subinterval $[0.3, 0.6)$, and a_4 is associated with the subinterval $[0.6, 1.0)$. Since the first symbol to be coded is a_4, the interval is narrowed to $[0.6, 1.0)$, which is then expanded to the full height. $[0.6, 1.0)$ is also subdivided into four regions based on the probability of each source symbol. Since the second symbol to be coded is a_3, the interval $[0.6, 1.0)$ is narrowed to $[0.72, 0.84)$, which is then expanded to the full height. The same process is continued until the last source symbol, which corresponds to a subinterval $[0.73272, 0.73344)$, is processed. Any number within this subinterval can be used to represent the whole sequence to be coded. In this example, 0.733 can be taken as the code for $a_4 a_3 a_2 a_1 a_3$.

One characteristic of arithmetic coding is that its decoding process can be performed with the help of its coding process. The input to the decoding process are the code word and the source information, A and u. The output from the decoding process is the sequence of symbols. Following the above example, the code word 0.733 should be located in the interval corresponding to a_4 as shown in Figure 6.3. This means

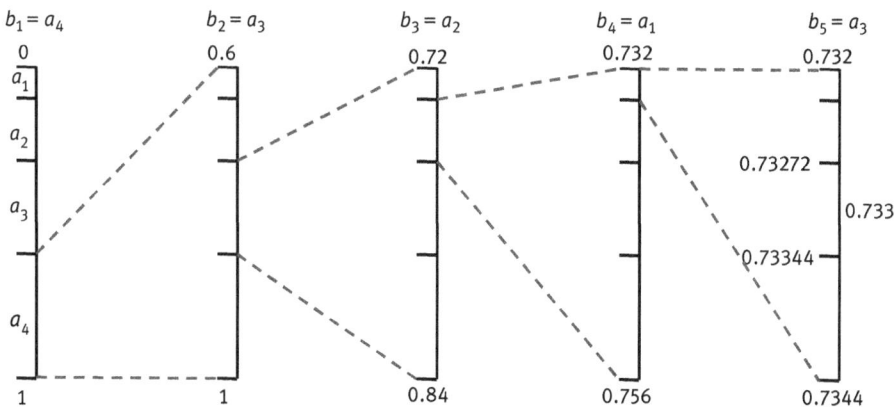

Figure 6.3: Arithmetic coding example.

that the first symbol in the decoding sequence is $b_1 = a_4$. Continuing this process by locating the intervals for 0.733 in the narrowed intervals gives $b_2 = a_3$, $b_3 = a_2$, $b_4 = a_1$, and $b_5 = a_3$. Note that this process can be continued, so in practice, a special end-of-sequence indicator should be added to the sequence.

Example 6.1 Arithmetic coding of a binary sequence in a binary system
Consider a binary source $A = \{a_1, a_2\} = \{0, 1\}$. The probabilities that the source will produce symbols a_1 and a_2 are $P(a_1) = 1/4$ and $P(a_2) = 3/4$, respectively. For a binary sequence 11111100, its arithmetic code word is 0.1101010_2. It is clear that the half-open interval [0, 1) in this case should be divided into 256 subintervals, so any binary sequence with eight symbols would be located in one of these subintervals.

◻

6.3 Bit-Plane Coding

Bit-plane coding decomposes one gray-level image into a sequence of bit-planes (binary images) and then codes each bit-plane. This technique can reduce/eliminate the coding redundancy as well as the inter-pixel redundancy. It has two main steps: bit-plane decomposition and bit-plane coding.

6.3.1 Bit-Plane Decomposition

There are several methods to decompose an image into a set of bit-planes. The mostly commonly used are binary code decomposition and Gray code decomposition.

6.3.1.1 Binary Code Decomposition
A gray-level image represented by a number of bits can be considered composed by the number of 1-bit-planes. For an 8-bit image, plane 0 contains all the lowest order bits of this image and plane 7 contains all the highest order bits of this image, as shown in Figure 6.4.

For an 8-bit gray-level image, when the value of plane 7 is 1, the gray-level values of its pixel must be greater than or equal to 128; when the value of plane 7 is 0, the gray-level values of its pixel must be less than or equal to 127.

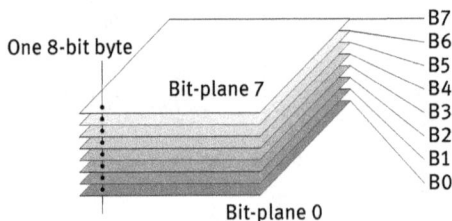

Figure 6.4: Binary code representation of an image and bit-plane.

Example 6.2 Illustration of the binary coded representation of bit-planes

Figure 6.5 gives a set of binary coded bit-planes of an 8-bit image. The images in Figure 6.5(a)–(h) are the eight bit-planes from plane 7 to plane 0. The high-order bit-planes are less complex than the low-order bit-planes, that is, they contain large uniform areas with significantly less detail or randomness. In Figure 6.5, only the first five high-order bit-planes have some visually meaningful information, while the other low-order bit-planes contain almost noisy pixels only. ◨

In general, the gray levels of an m-bit image can be represented in the form of a base 2 polynomial

$$a_{m-1}2^{m-1} + a_{m-2}2^{m-2} + \cdots + a_1 2^1 + a_0 2^0 \tag{6.16}$$

The decomposition of the image into a set of bit-planes is to separate the m coefficients of the polynomial into m 1-bit bit-planes. The disadvantage of this approach is that small changes in the gray levels can have a significant impact on the complexity of the bit-planes. For example, if a pixel of a gray-level value of 127 (01111111_2) is adjacent to a pixel of gray-level value of 128 (10000000_2), every bit-plane will contain a transition from 0 to 1 (or 1 to 0).

6.3.1.2 Gray Code Decomposition

To reduce the effect of small gray-level variations on a bit-plane, an alternative decomposition approach can be used. It represents the image first by an m-bit **Gray code**.

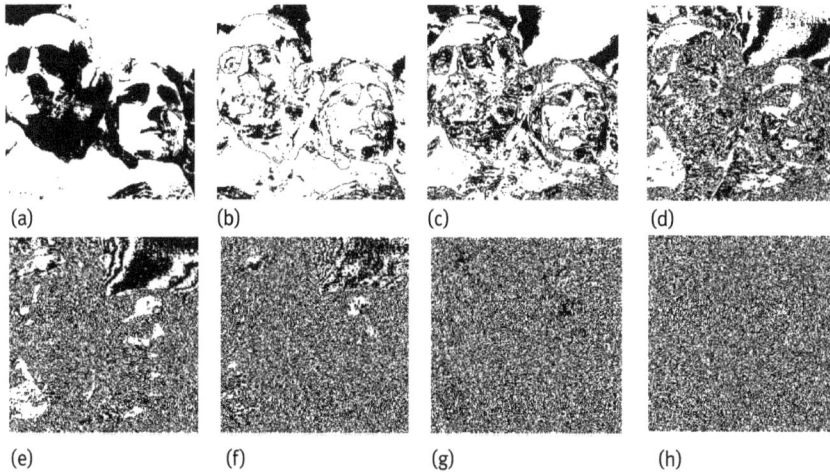

(a)   (b)   (c)   (d)

(e)   (f)   (g)   (h)

Figure 6.5: Examples of binary coded representation of bit-plane.

The m-bit Gray code corresponding to the polynomial in eq. (6.16) can be computed from ($\oplus$ denotes the exclusive-OR operation)

$$g_i = \begin{cases} a_i \oplus a_{i+1} & 0 \leq i \leq m - 2 \\ a_i & i = m - 1 \end{cases} \tag{6.17}$$

This code has the unique property that successive code words differ in only one bit position. Thus, small changes in gray levels are less likely to affect all m-bit-planes. For example, when gray levels 127 and 128 are adjacent, only the seventh bit-plane will contain a 0 to 1 transition, since the Gray codes that correspond to 127 and 128 are 01000000_2 and 11000000_2.

Example 6.3 Illustration of the Gray-coded representation of bit-planes
Figure 6.6 gives a set of Gray-coded bit-planes of an 8-bit image, which is also used in Example 6.2. The images in Figure 6.6(a)–(h) are the 8-bit-planes from plane 7 to plane 0. The high-order bit-planes are still less complex than the low-order bit-planes, as in Figure 6.5. Comparing Figure 6.6 with Figure 6.5, the complexity of the Gray-coded bit-planes is less than that of the binary coded bit-planes. In other words, the number of bit-planes with meaningful visual information in the Gray-coded representation is greater than in the binary coded representation. ▫

6.3.2 Bit-Plane Coding

The results of bit-plane decomposition are binary images with values 0 and 1. As shown in Figures 6.5 and 6.6, there are many connected (constant) regions in

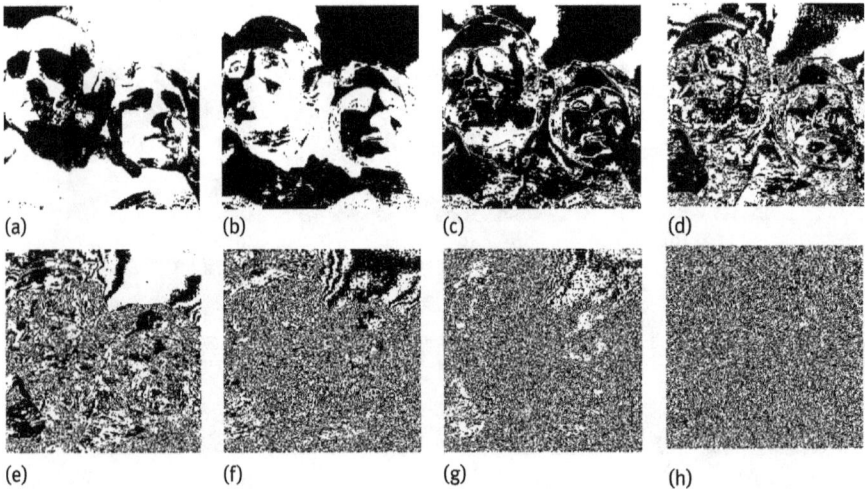

(a)　　　　　(b)　　　　　(c)　　　　　(d)

(e)　　　　　(f)　　　　　(g)　　　　　(h)

Figure 6.6: Examples of Gray-coded representation of bit-planes.

bit-planes. In addition, pixels with value 0 and pixels with value 1 are complementary in bit-planes. All these properties can be used to help the coding of images.

6.3.2.1 Constant Area Coding

For a bit-plane, using a special code word to represent each connected component is very effective. **Constant area coding** (CAC) is a simple method. It divides the image into blocks of size $m \times n$ pixels, which can be all white, all black, or mixed. The most probable occurring type is assigned the 1-bit code word 0, and the other two types are assigned the 2-bit code words 10 and 11. Compression is achieved because each of the mn bits that normally would be used to represent each constant area is replaced by either 1-bit or 2-bit code words. Note that the code assigned to the mixed type is used as a prefix, which is followed by the mn-bit pattern of the block.

In the case where the image to be compressed is dominant with white blocks (such as text documents), the white blocks can be coded with 0 and the rest of the blocks (including black blocks) can be coded with a 1 followed by the bit pattern of the block. Such an approach is called **white block skipping** (WBS), and is simpler than the standard CAC. As few black blocks are expected, they are grouped with the mixed blocks. An even more effective modification with blocks of size $1 \times n$ is to code the white lines as 0s and all other lines with a 1 followed by the normal WBS code sequence.

6.3.2.2 1-D Run-Length Coding

The **1-D Run-length coding** technique represents each row of an image by a sequence of run-lengths that describe successive runs of black and white pixels. The basic concept is to code each contiguous group of 0s or 1s encountered in a left-to-right scan of a row by its run-length. This requires establishing conventions to determine the run-length. The most common conventions are: (1) the value of the first run of each row must be specified and (2) each row must begin with a white run, whose run-length can be zero.

By coding the length of each run with variable-length coding, additional compression can be achieved. The black-and-white run-lengths can be coded separately by counting their own statistics. For example, letting symbol a_j represent a black run of length j, the probability that symbol a_j was emitted by an imaginary black run-length source can be estimated by dividing the number of the black run-lengths of length j in the entire image by the total number of black runs.

6.3.2.3 2-D Run-Length Coding

The above 1-D coding concept can be extended to **2-D run-length coding**. One of the better known results is the relative address coding (RAC), which is based on the principle of tracking the binary transitions between the beginning and end points of each run. Figure 6.7 illustrates one implementation of this approach. Note that ec is the distance from the current transition c to the last transition of the current line e, whereas

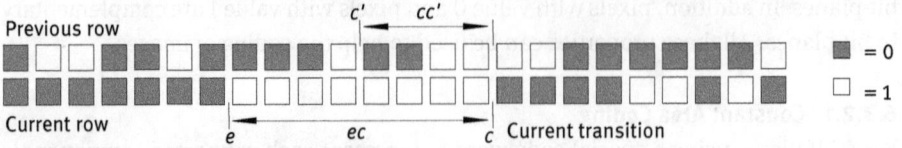

Figure 6.7: Illustration of RAC distance.

cc' is the distance from c to the first similar transition passing e in the same direction, denoted c', on the previous line. If $ec \leq cc'$, the RAC coded distance d is set to ec and used to represent the current transition at c. If $cc' < ec$, d is set to cc'.

RAC requires the adoption of a convention for determining run values. In addition, imaginary transitions at the beginning and end points of each line, as well as an imaginary starting line must be assured so that the image boundaries can be handled properly. Finally, since the probability distributions of the RAC distances of natural images are not uniform, the final step of the RAC process is to code the RAC distance measured by using a suitable variable-length code.

Example 6.4 Coding RAC distance with variable-length code

Table 6.6 presents a method for RAC distance coding. The shortest distance should be coded with the shortest code. All other distances are coded as follows. The first prefix is used to present the shortest RAC distance. The second prefix is used to assign d to a specific range of the distance, and it is followed by the binary representation (denoted #...# in Table 6.6) of d minus the base distance of the range itself. In Figure 6.7, ec and cc' equals +8 and +4, respectively, so the proper RAC code word is 1100011. Since cc' is +4 and ec is +8, cc' is taken first, and the first prefix is 1100. The second prefix code $h(d)$ should be in the form of 0## (RAC distance is 4). Further, ## should be 11 (d minus the base distance is $4 - 1$). Finally, if $d = 0$, c is directly below c', whereas if $d = 1$, the decoder may have to determine the closest transition point, because the code 100 does not specify whether the measurement is relative to the current row or to the previous row. ◻

Table 6.6: Coding RAC distance with a variable-length code.

Distance measured	RAC Distance	Prefix code 1	Distance range	Prefix code 2 $h(d)$
cc'	0	0	1 ~ 4	0 ##
ec or cc' (left)	1	100	5 ~ 20	10 ####
cc' (right)	1	101	21 ~ 84	110 ######
ec	$d(d > 1)$	111 $h(d)$	85 ~ 340	1110 ########
cc' (c' to left)	$d(d > 1)$	1100 $h(d)$	341 ~ 1,364	11110 #########
cc' (c' to right)	$d(d > 1)$	1101 $h(d)$	1,365 ~ 5,460	111110 ##########

Table 6.7: A comparison of three techniques ($H \approx 6.82$ bits/pixel)

Bit-plane coding	Method	Bit-plane code rate (bits/pixel)								Code rate	Compression ratio
		7	6	5	4	3	2	1	0		
	CAC (4 × 4)	0.14	0.24	0.60	0.79	0.99	—	—	—	5.75	1.4 : 1
Binary	RLC	0.09	0.19	0.51	0.68	0.87	1.00	1.00	1.00	5.33	1.5 : 1
	RAC	0.06	0.15	0.62	0.91	—	—	—	—	5.74	1.4 : 1
	CAC (4 × 4)	0.14	0.18	0.48	0.40	0.61	0.98	—	—	4.80	1.7 : 1
Gray	RLC	0.09	0.13	0.40	0.33	0.51	0.85	1.00	1.00	4.29	1.9 : 1
	RAC	0.06	0.10	0.49	0.31	0.62	—	—	—	4.57	1.8 : 1

6.3.2.4 Comparison

A comparison of the above three techniques for the coding of a gray-level image gives the results shown in Table 6.7 (Gonzalez and Woods, 1992), in which the first-order estimates of the entropies of the RLC run-lengths (H) are used as an approximation of the compression performance that could be achieved under the variable-length coding approaches.

From Table 6.7, it is noted that:

(1) All three techniques can eliminate some amount of inter-pixel redundancy, as all resulting code rates are less than H. Among them, RLC is the best.
(2) The performance of Gray coding is over the performance of the binary coding by about 1 bit/pixel.
(3) All techniques can compress the monochrome image only by a factor of 1 to 2, as they cannot successfully compress the lower order bit-planes.

6.4 Predictive Coding

The basic idea behind **predictive coding** is to reduce or to remove inter-pixel redundancy by only coding the new information in each pixel. Here, the new information in a pixel is defined as the difference between the current value of the pixel and the predictive value of the pixel. Note that the existence of the correlation between pixels in an image makes the prediction possible.

Predictive coding can be either lossless or lossy. They will be discussed separately in the following sections.

6.4.1 Lossless Predictive Coding

A **lossless predictive coding system** is composed of an encoder and a decoder, as shown in Figure 6.8.

Both the encoder and the decoder have an identical predictor. When the pixels of the input image $f_n (n = 1, 2, \ldots)$ are added into the encoder serially, the predictor

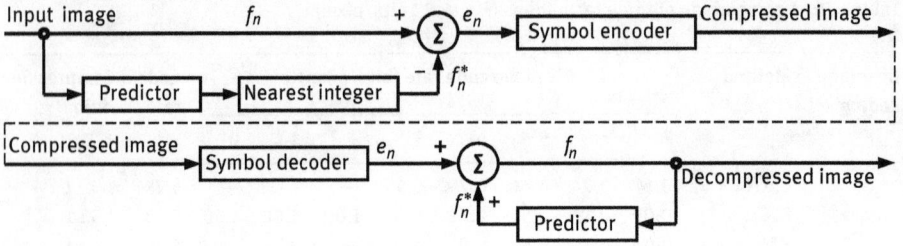

Figure 6.8: A system of lossless predictive coding.

produces the predictive/estimation value of the current pixel according to the values of several previous input pixels. The output of the predictor is then rounded to the nearest integer, denoted f_n^*, and used to form the prediction error

$$e_n = f_n - f_n^*$$ (6.18)

This error will be coded by the symbol encoder with a variable-length code to generate the element of the compressed image. On the other side, the decoder will reconstruct e_n with the received variable-length code and perform the following operation to obtain the original image

$$f_n = e_n + f_n^*$$ (6.19)

In many cases, the prediction is formed by a linear combination of m previous pixels as follows

$$f_n^* = \text{round}\left[\sum_{i=1}^{m} a_i f_{n-i}\right]$$ (6.20)

where m is called the order of the linear predictor, round $[\cdot]$ is a rounding function, and a_i for $i = 1, 2, \ldots, m$ are the prediction coefficients. The subscript n in eqs. (6.18)–(6.20) indicates the spatial coordinates. For 1-D linear predictive coding, eq. (6.21) can be written as

$$f_n^*(x, y) = \text{round}\left[\sum_{i=1}^{m} a_i f(x, y - i)\right]$$ (6.21)

According to eq. (6.21), the 1-D linear prediction $f^*(x, y)$ is a function of the previous pixels on the current line. In a 2-D linear prediction $f^*(x, y)$ is a function of the previous pixels in a left-to-right, top-to-bottom scan of an image.

The simplest 1-D linear predictive coding corresponds to $m = 1$, that is,

$$f_n^*(x, y) = \text{round}\left[a f(x, y - 1)\right]$$ (6.22)

The predictor represented by eq. (6.22) is also called a previous pixel predictor and the corresponding predictive coding is also called differential coding or previous pixel coding.

6.4.2 Lossy Predictive Coding

Adding a quantizer into the system of lossless predictive coding as shown in Figure 6.8, a system of the **lossy predictive coding** can be formed, as shown in Figure 6.9. The quantizer is inserted between the symbol encoder and the point at which the predictive error is formed. It absorbs the nearest integer function of the error-free encoder. It maps the prediction error into a limited range of outputs, denoted e'_n, which establishes the amount of compression and distortion associated with the lossy predictive coding.

To accommodate the insertion of the quantization step, the lossless encoder shown in Figure 6.8 must be changed so that the predictions generated by the encoder and decoder are equivalent. In Figure 6.9, the predictor is placed in a feedback loop. The input of the predicator is generated as a function of past predications and the corresponding quantized errors. This is given by

$$f'_n = \dot{e}_n + f^*_n \tag{6.23}$$

Such a closed loop prevents the error from being built up in the decoder's output, which is also given by eq. (6.23).

A simple and effective form of the lossy predictive coding has the predictor and quantizer that are defined as

$$f^*_n = a f'_{n-1} \tag{6.24}$$

$$e'_n = \begin{cases} +c & e_n > 0 \\ -c & \text{otherwise} \end{cases} \tag{6.25}$$

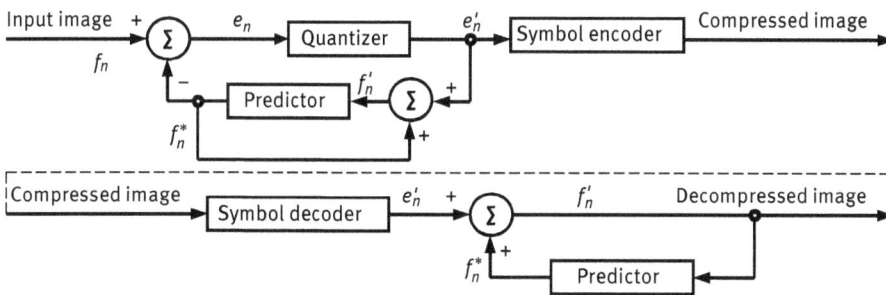

Figure 6.9: A system of lossy predictive coding.

where a is a prediction coefficient (normally less than 1) and c is a positive constant. The output of the quantizer can be represented by a single bit, so the symbol encoder of Figure 6.9 can use a 1-bit fixed-length code. The resulting **Delta modulation** (DM) code rate is 1 bit/pixel.

Example 6.5 DM coding example

A DM coding example is given in Table 6.8. Suppose the input sequences are {12, 16, 12, 14, 20, 32, 46, 52, 50, 51, 50}. Let $a = 1$ and $c = 5$.

The coding process starts with the error-free transfer of the first input pixel to the decoder. With the initial condition $f_0' = f_0 = 12.0$ established for both the encoder and

Table 6.8: A DM coding example.

Input		Encoder				Decoder		Error
N	f	f*	E	e'	f'	f*	f'	[f − f']
	12.0	−	−	−	12.0	−	12.0	0.0
1	16.0	12.0	4.0	5.0	17.0	12.0	17.0	−1.0
2	14.0	17.0	−3.0	−5.0	12.0	17.0	12.0	2.0
3	18.0	12.0	6.0	5.0	17.0	12.0	17.0	1.0
4	22.0	17.0	5.0	5.0	22.0	17.0	22.0	0.0
5	32.0	22.0	10.0	5.0	27.0	22.0	27.0	5.0
6	46.0	27.0	19.0	5.0	32.0	27.0	32.0	14.0
7	52.0	32.0	20.0	5.0	37.0	32.0	37.0	15.0
8	50.0	37.0	8.0	5.0	42.0	37.0	42.0	8.0
9	51.0	42.0	9.0	5.0	47.0	42.0	47.0	4.0
10	50.0	47.0	3.0	5.0	52.0	47.0	52.0	−2.0

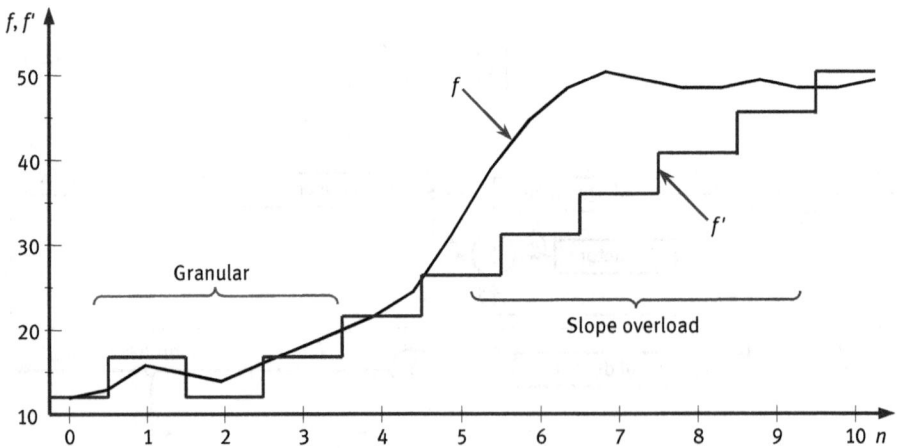

Figure 6.10: Distortions in DM coding.

the decoder, the remaining outputs can be computed by repeatedly using eqs. (6.24), (6.18), (6.25), and (6.23).

Figure 6.10 shows the input and the output of Table 6.8. Two things need to be pointed out. One is when c is too large compared to the variation of input, in the relatively smooth region from $n = 0$ to $n = 3$, DM coding will produce granular noise. Another is when c is too small to represent the input's largest changes, in the cliffy region from $n = 5$ to $n = 9$, DM coding will produce slope overload.   ◙

There are interactions between the predictor and the quantizer in the encoder. However, the predictor and the quantizer are often designed independently. That is, the predictor is designed under the assumption that no quantization error exists, and the quantizer is designed to minimize its own error.

6.4.2.1 Optimal Predictors

In most predictive coding applications, the encoder's mean-square prediction error is minimized to achieve the optimal performance for **optimal predictors**. This is given by

$$E\left\{e_n^2\right\} = E\left\{[f_n - f_n^*]^2\right\} \tag{6.26}$$

subject to the constrain that

$$f_n' = e_n' + f_n^* \approx e_n + f_n^* = f_n \tag{6.27}$$

$$f_n^* = \sum_{i=1}^{m} a_i f_{n-i} \tag{6.28}$$

Suppose that the quantization error is negligible ($e_n' \approx e_n$), and the prediction is constrained to a linear combination of the m previous pixels, then the predictive coding method is called the **differential pulse code modulation** (DPCM). In this approach, the m prediction coefficients should be determined to minimize the following expression

$$E\left\{e_n^2\right\} = E\left\{\left[f_n - \sum_{i=1}^{m} a_i f_{n-i}\right]^2\right\} \tag{6.29}$$

It has been proven that for a 2-D Markov source, a generalized fourth-order linear predictor can be written as follows

$$f^*(x, y) = a_1 f(x, y - 1) + a_2 f(x - 1, y - 1) + a_3 f(x - 1, y) + a_4 f(x + 1, y - 1) \tag{6.30}$$

By assigning different values to the coefficients in eq. (6.30), different predictors can be formed. Some examples are as follows:

$$f_1^*(x, y) = 0.97f(x, y-1) \tag{6.31}$$

$$f_2^*(x, y) = 0.5f(x, y-1) + 0.5f(x-1, y) \tag{6.32}$$

$$f_3^*(x, y) = 0.75f(x, y-1) + 0.75f(x-1, y) - 0.5f(x-1, y-1) \tag{6.33}$$

$$f_4^*(x, y) = \begin{cases} 0.97f(x, y-1) & |f(x-1, y) - f(x-1, y-1)| \le |f(x, y-1) - f(x-1, y-1)| \\ 0.97f(x-1, y) & \text{otherwise} \end{cases} \tag{6.34}$$

Equation (6.34) is an adaptive predictor.

The sum of the prediction coefficients in eq. (6.28) normally is required to be less than or equal to 1. That is,

$$\sum_{i=1}^{m} a_i \le 1 \tag{6.35}$$

Example 6.6 A comparison of different predictors

The decompressed images of Figure 1.1(a), using the predictors of eqs. (6.31)–(6.34) after compression, are shown in Figures 6.11(a)–(d). For all predictors, the two-level quantizer as given by eq. (6.25) is used.

It can be seen that the visual quality of decompressed images increases with the order of the predictors. Figure 6.11(c) is better than Figures 6.11(b) and (b) is better than Figure 6.11(a). Figure 6.11(d) corresponds to the adaptive predictor of order 1. Its quality

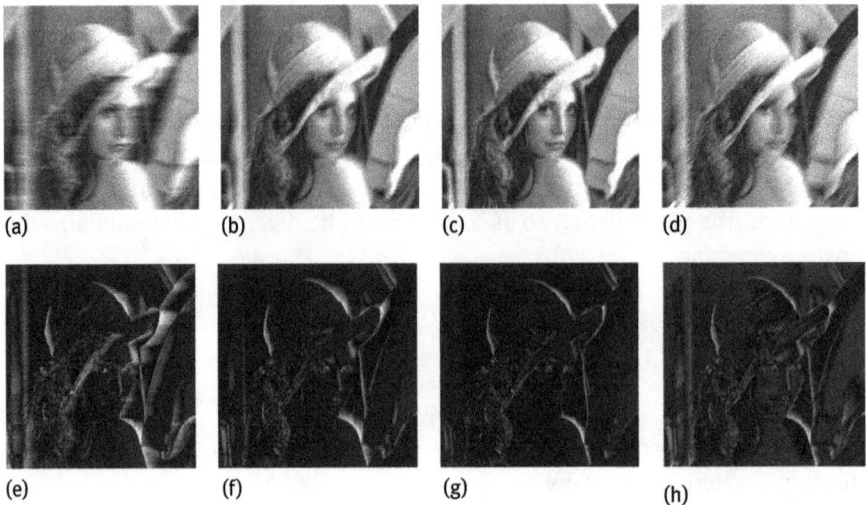

(a)   (b)   (c)   (d)

(e)   (f)   (g)   (h)

Figure 6.11: Comparison of predictors.

is better than Figure 6.11(a) but worse than Figure 6.11(b). A similar conclusion can also be made from Figure 6.11(e)–(h), which are the corresponding error images (showing the difference between original images and decompressed images). ◻

6.4.2.2 Optimal Quantization

A typical quantization function is shown in Figure 6.12. This staircase function $t = q(s)$ is an odd function of s. It can be completely described by the $L/2$ s_i (decision levels) and t_i (reconstruction levels) in the first quadrant. As a convention, s in the half-open interval $(s_i, s_{i+1}]$ is mapped to t_{i+1}.

In the design of a quantizer, the best s_i and t_i for a given optimization criterion and the input probability density function $p(s)$ should be selected. If using the mean-square quantization error $E\{(s - t_i)^2\}$ as the criterion for **optimal quantization**, and $p(s)$ is an even function, the condition for the minimal error is

$$\int_{s_{i-1}}^{s_i} (s - t_i)p(s)ds = 0 \quad i = 1, 2, \cdots, L/2 \tag{6.36}$$

where

$$s_i = \begin{cases} 0 & i = 0 \\ (t_i + t_{i+1})/2 & i = 1, 2, \cdots, L/2 - 1 \\ \infty & i = L/2 \end{cases} \tag{6.37}$$

$$s_i = -s_{-i} \qquad t_i = -t_{-i} \tag{6.38}$$

Each equation has a particular meaning. Equation (6.36) indicates that the recon-struction levels are the centroids of the areas under $p(s)$ over the specified decision intervals. Equation (6.37) indicates that the decision levels are halfway between the reconstruction levels. For any L, the s_i and t_i satisfy eqs. (6.36)–(6.38) and are optimal in the mean-square error sense; the corresponding quantizer is called an L-level Lloyd-Max quantizer.

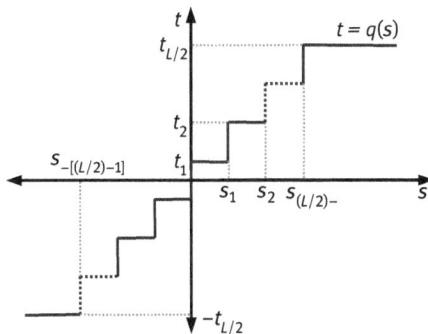

Figure 6.12: A typical quantization function.

Table 6.9: Performances of different combinations of predictors and quantizers.

Predictor	Lloyd-Max quantizer			Adaptive quantizer		
	2 levels	4 levels	8 levels	2 levels	4 levels	8 levels
Equation (6.31)	30.88	6.86	4.08	7.49	3.22	1.55
Equation (6.32)	14.59	6.94	4.09	7.53	2.49	1.12
Equation (6.33)	8.90	4.30	2.31	4.61	1.70	0.76
Equation (6.34)	38.18	9.25	3.36	11.46	2.56	1.14
Compression ratio	8.00 : 1	4.00 : 1	2.70 : 1	7.11 : 1	3.77 : 1	2.56 : 1

(a)   (b)   (c)   (d)   (e)

Figure 6.13: A comparison of different quantizers in DPCM coding.

Table 6.9 shows the mean-square errors for the different combinations of predictors and quantizers. It can be seen that the two-level adaptive quantizers have similar performance as the four-level nonadaptive quantizers, and the four-level adaptive quantizers perform similar to the eight-level nonadaptive quantizers.

Example 6.7 A comparison of different quantizers in DPCM coding
Figures 6.13(a)–(c) show decompressed images with quantizers of 5, 9, and 17 levels (the predicator used is order 1 as shown in eq. (6.31)). Figures 6.13(d) and (e) show the error images corresponding to Figures 6.13(a) and (b). In fact, when the quantization level is 9 the errors are hardly perceptible, and even higher quantization levels are useless. ◻

6.5 Transform Coding

Predictive coding techniques directly operate in the image space and are called spatial domain methods. Coding can also be made in the frequency domain, such as the transform coding technique presented in this section. In general, transform coding is lossy.

6.5.1 Transform Coding Systems

A typical **transform coding system** is shown in Figure 6.14. The coder consists of four operation modules: sub-image decomposition, transformation, quantization,

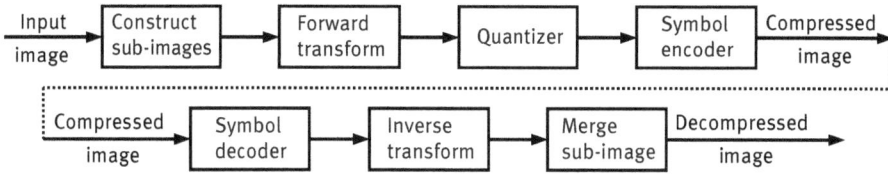

Figure 6.14: A typical transform coding system.

and symbol coding. An $N \times N$ image is first divided into sub-images of size $n \times n$, which are then transformed to generate $(N/n)^2$ $n \times n$ sub-image transform arrays. The purpose of the transformation is to de-correlate the pixels of each sub-image, or to pack as much information as possible into the smallest number of the transform coefficients. In the following quantization stage, the coefficients carrying the least information will be eliminated or more coarsely quantized. The final process is to code (often with a variable-length code, as discussed in Section 6.2) the quantized coefficients. The decoder consists of an inverse sequence of modules, with the exception of the quantization step, which cannot be inversed.

Since a certain number of data are lost in the quantization stage, transform coding techniques are always lossy. Any or all of the transform encoding steps can be adapted to the local image content, which is called adaptive transform coding, or fixed for all sub-images, which is called nonadaptive transform coding.

6.5.2 Sub-Image Size Selection

Sub-image decomposition is used for reducing the correlation between adjacent sub-images. The sub-image size is a critical factor, as both the compression ratio and the computation effort are related (propositional) to this size. Two factors to be considered in the selection of this size are as follows: The correlation between adjacent sub-images should be reduced to an acceptable level and the length and the width of the sub-image should be an integer power of 2.

Example 6.8 The influence of the sub-image size
A graphic illustration of the influence of the sub-image size on the reconstruction error is given in Figure 6.15. Three transforms are considered. They are the Fourier transform, the Hadamard transform, and the cosine transform. The horizontal axis represents the size of sub-image n with $n = 2, 4, 8, 16, 32$. The vertical axis represents the root-mean-square error (RMSE). The quantization is performed by truncating 75% of the transform coefficients.

It can be seen that the curves corresponding to both the Hadamard transform and the cosine transform are flat when the sub-image size is bigger than 8×8, while the curve corresponding to the Fourier transform decreases quickly in this region.

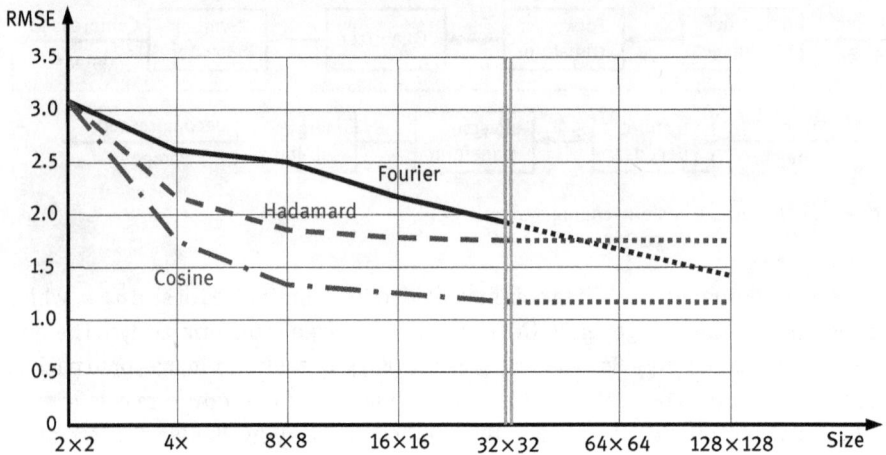

Figure 6.15: Reconstruction error versus sub-image size.

Extending these curves to a size larger than 32×32 (in the right side of the double line), the Fourier curve will cross through the Hadamard curve and approach (converge) the cosine curve.

◻

6.5.3 Transform Selection

In transform coding, compression is achieved during the quantization step but not the transform step. Various transforms can be used in transform coding. For a given coding application, the selection of the transform depends on the error allowed and the computational complexity.

6.5.3.1 Mean-Square Error of Reconstruction

An $N \times N$ image $f(x,y)$ can be represented as a function of its 2-D transform $T(u,v)$

$$f(x,y) = \sum_{u=0}^{n-1}\sum_{v=0}^{n-1} T(u,v)h(x,y,u,v) \qquad x,y = 0,1,\cdots,n-1 \qquad (6.39)$$

Here, n is used to present the size of sub-image. The inverse kernel $h(x,y,u,v)$ depends only on x, y, u, and v, but not on $f(x,y)$ or $T(u,v)$, and it can be considered a set of basis images. Equation (6.39) can be represented also by

$$F = \sum_{u=0}^{n-1}\sum_{v=0}^{n-1} T(u,v)H_{uv} \qquad (6.40)$$

where F is an $n \times n$ matrix containing the pixels of $f(x, y)$, and

$$
H_{uv} = \begin{bmatrix} h(0, 0, u, v) & h(0, 1, u, v) & \cdots & h(0, n-1, u, v) \\ h(1, 0, u, v) & \vdots & \cdots & \vdots \\ \vdots & \vdots & \ddots & \vdots \\ h(n-1, 0, u, v) & h(n-1, 1, u, v) & \cdots & h(n-1, n-1, u, v) \end{bmatrix} \tag{6.41}
$$

Equation (6.40) defined F as a linear combination of n^2 $n \times n$ matrices (H_{uv}). These matrices can be used to compute $T(u, v)$.

Now, define a transform coefficient masking function

$$
m(u, v) = \begin{cases} 0 & \text{if } T(u, v) \text{ satisfies a specified truncation criterion} \\ 1 & \text{otherwise} \end{cases} \tag{6.42}
$$

An approximation of F can be obtained from the truncated expansion

$$
F^* = \sum_{u=0}^{n-1} \sum_{v=0}^{n-1} T(u, v)m(u, v)H_{uv} \tag{6.43}
$$

The mean-square error between the sub-image F and its approximation F^* is

$$
e_{ms} = E\left\{\|F - F^*\|^2\right\} = E\left\{\left\|\sum_{u=0}^{n-1}\sum_{v=0}^{n-1} T(u, v)H_{uv}\left[1 - m(u, v)\right]\right\|^2\right\}
$$
$$
= \sum_{u=0}^{n-1}\sum_{v=0}^{n-1} \sigma^2_{T(u,v)}H_{uv}\left[1 - m(u, v)\right] \tag{6.44}
$$

where $\|F - F^*\|$ is the matrix form of $(F - F^*)$ and $\sigma^2_{T(u,v)}$ is the variance of the coefficient at transform location (u, v). The mean-square errors of the $(N/n)^2$ sub-images of an $N \times N$ image are identical, so the mean-square error of the $N \times N$ image equals the mean-square error of a single sub-image.

The total mean-square approximation error between the original image and the decompressed image is the sum of the variance of the discarded transform coefficients. The performance of the transform coding is thus related to the variance distribution of the transform coefficients. Transformations that redistribute or pack the most information into the fewest coefficients provide the best sub-image approximation and the smallest reconstruction error.

6.5.3.2 Discussions on Transform Selection
A comparative discussion concerning DFT, WHT, DCT, and KLT follows.

From the point of view of information packing ability, KLT is the best transform. The KLT can minimize the mean-square error of eq. (6.44) for any image and any number of retained coefficients. KLT is followed by DFT and DCT, while WHT is the worst among all four transforms.

From the perspective of computation, WHT is the simplest as only 1 and −1 are involved in the computation. WHT is followed by DFT and DCT. Both of them are independent of the input image and have fixed basic kernel functions. The basis functions of KLT are related to the input image (data dependent), so the computation of the basis function must be carried out for all sub-images. For this reason, KLT needs a lot of computation and thus is rarely used in practice. On the other hand, transforms such as DFT, WHT, and DCT, whose basis images are fixed, are commonly used.

Many transform coding systems currently used are based on DCT, which provides a good compromise between information packing ability and computational complexity. One advantage is that DCT can pack the most information into the fewest coefficients for the most natural images. Another advantage is its ability to minimize the block-like appearance, called the blocking artifact. This is because that DCT has a 2N-point periodicity (for an N point transform), which does not inherently produce boundary discontinuities.

6.5.4 Bit Allocation

The reconstruction error associated with the truncated series expansion of eq. (6.43) is a function of the number and relative importance of the transform coefficients that are discarded, as well as the precision that is used to represent the retained coefficients. Two criteria can be used: maximum variance or maximum magnitude. The former is called zonal coding and the latter is called threshold coding. The overall process of truncating, quantizing, and coding the coefficients of a transformed sub-image is called **bit allocation.**

6.5.4.1 Zonal Coding
Zonal coding is based on the principle of information theory that views information as uncertain. The coefficients with maximum variance would carry the most information and should be retained. The variances can be calculated directly from the ensemble of $(N/n)^2$ transformed sub-images. According to eq. (6.43), the zonal sampling process is used to multiply each $T(u, v)$ by the corresponding element in a zonal mask, which is formed by placing a 1 in the locations of the maximum variance and a 0 in all other locations. Usually, coefficients of the maximum variance are located around the origin of the transformed space (left-top), resulting in the typical zonal mask shown in Figure 6.16(a), in which the coefficients retained are highlighted by the shading.

The coefficients retained during the zonal sampling process must be quantized and coded, so zonal masks are sometimes depicted by the number of bits used, as shown in Figure 6.16(b). Two strategies used are as follows: the same numbers of bits

1	1	1	1	1	0	0	0
1	1	1	1	0	0	0	0
1	1	1	0	0	0	0	0
1	1	0	0	0	0	0	0
1	0	0	0	0	0	0	0
0	0	0	0	0	0	0	0
0	0	0	0	0	0	0	0
0	0	0	0	0	0	0	0

(a)

8	7	6	4	3	2	1	0
7	6	5	4	3	2	1	0
6	5	4	3	3	1	1	0
4	4	3	3	2	1	0	0
3	3	3	2	1	1	0	0
2	2	1	1	1	0	0	0
1	1	1	0	0	0	0	0
0	0	0	0	0	0	0	0

(b)

Figure 6.16: Typical zonal masks and zonal bit allocation.

are allocated for different coefficients and some fixed number of bits are distributed among them unequally. In the first case, the coefficients are normalized by their standard deviations and uniformly quantized. In the second case, a quantizer is designed for each coefficient.

6.5.4.2 Threshold Coding
In zonal coding, a single fixed mask is used for all sub-images. In **threshold coding**, the location of the transform coefficients retained for each sub-image varies. The underlying idea is that, for any sub-image, the transform coefficients of the largest magnitude make the most significant contribution to reconstructed sub-image quality. Since the locations of the maximum coefficients vary from one sub-image to another, the elements of $T(u, v)$ and $m(u, v)$ normally are reordered to form a 1-D, run-length coded sequence. One typical threshold mask is shown in Figure 6.17(a). When the mask is applied, via eq. (6.43), to the sub-image for which it was derived, and the resulting $n \times n$ array is reordered to form a n^2-element coefficient sequence in accordance with the zigzag ordering pattern (from 0 to 63) of Figure 6.17(b), the reordered 1-D sequence contains several long runs of 0s. These runs can be run-length coded. The

1	1	1	1	0	1	0	0
1	1	1	1	0	0	0	0
0	1	0	0	0	0	0	0
1	0	0	0	0	0	0	0
0	0	0	0	0	0	0	0
0	1	0	0	0	0	0	0
0	0	0	0	0	0	0	0
0	0	0	0	0	0	0	0

(a)

0→1	5→6	14	15	27	28
2	4	7	13	16	26	29	42
3	8	12	17	25	30	41	43
9	11	18	24	31	40	44	53
10	19	23	32	39	45	52	54
20	22	33	38	46	51	55	60
21	34	37	47	50	56	59	61
35	36	48	49	57	58	62→63	

(b)

Figure 6.17: Typical threshold masks and thresholded coefficient ordering sequence.

nonzero or retained coefficients, corresponding to the mask locations that contain a 1, can be represented using variable-length codes.

There are three basic ways to threshold a transformed sub-image. They are as follows:

(1) Using a single global threshold for all sub-images, the level of compression differs from image to image, depending on the number of coefficients that exceed the global threshold.

(2) Using a different threshold for each sub-image, the same number of coefficients is discarded for each sub-image, which is referred to as N-largest coding.

(3) Using a varying threshold, which is a function of the location of each coefficient, a variable code rate is obtained, offering the advantage that thresholding and quantization can be combined by replacing $T(u, v)$ and $m(u, v)$ with

$$T_N(u, v) = \text{round} \left[\frac{T(u, v)}{N(u, v)} \right] \qquad (6.45)$$

where $T_N(u, v)$ is a thresholded and quantized approximation of $T(u, v)$, and $N(u, v)$ is an element of the transform normalization array

$$\mathbf{N} = [N(u, v)] = \begin{bmatrix} N(0, 0) & N(0, 1) & \cdots & N(0, n-1) \\ N(1, 0) & \vdots & \cdots & \vdots \\ \vdots & \vdots & \ddots & \vdots \\ N(n-1, 0) & N(n-1, 1) & \cdots & N(n-1, n-1) \end{bmatrix} \qquad (6.46)$$

Before a normalized (thresholded and quantized) sub-image transform, $T_N(u, v)$, can be inverse transformed to obtain an approximation of the sub-image $f(x, y)$, it must be multiplied by $N(u, v)$. The resulting denormalized image, denoted $T_A(u, v)$, is an approximation of $T_N(u, v)$

$$T_A(u, v) = T_N(u, v)N(u, v) \qquad (6.47)$$

The inverse transformation of $T_A(u, v)$ yields the decompressed sub-image approximation.

Figure 6.18 shows the quantation curve when $N(u, v)$ is a constant c. $T_N(u, v)$ assumes the integer value k if and only if $kc - c/2 \leq T(u, v) < kc + c/2$. If $N(u, v) > 2T(u, v)$, then $T_N(u, v) = 0$ and the transform coefficient is completely truncated or discarded.

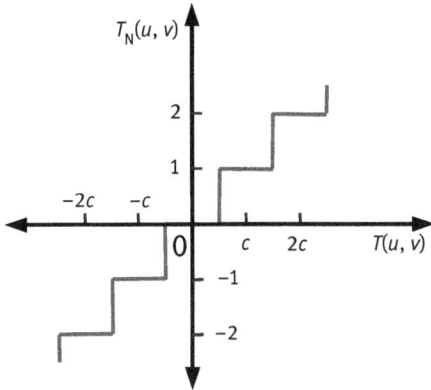

Figure 6.18: Illustration of threshold coding quantization curve.

6.5.5 Wavelet Transform Coding

Wavelet transform coding is also a transform coding method based on reducing the correlation among pixels to obtain the compression effect. A typical **wavelet transform coding system** is illustrated in Figure 6.19.

Comparing Figures 6.19 with 6.14, it is clear that there is no module for sub-image decomposition in the wavelet transform coding system. This is because the wavelet transform is efficient in computation and has inherent local characteristics.

In the following, some influence factors in wavelet transform coding are discussed.

6.5.5.1 Wavelet Selection
The selection of wavelet affects the design of the wavelet transform coding system. The performance of the system, such as the computational complexity and the error produced, is closely related to the type of wavelet. The most popularly used wavelets include the Haar wavelet, the Daubechies wavelet, and the bi-orthogonal wavelet.

6.5.5.2 Decomposition Level Selection
The number of decomposition levels has both an impact on computational complexity and reconstruction error. A fast wavelet transform of P scales includes the iteration of P filter banks, so the number of operations for the wavelet transforms and inverse

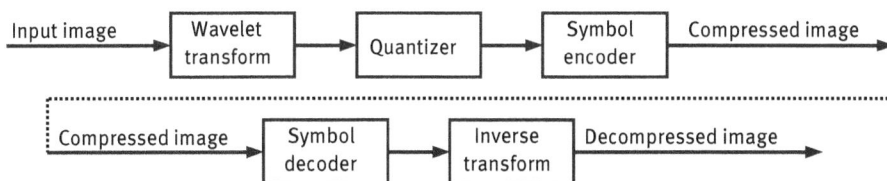

Figure 6.19: A typical coding system with wavelet transforms.

wavelet transforms would increase with the rise of decomposition levels. In addition, with the increase of the decomposition levels, the quantization for low scale coefficients will increase, too. This will influence larger regions in the reconstructed image. In many applications, a compromise between the resolution of the image and the lowest scale of approximation should be made.

6.5.5.3 Quantization
The largest factor influencing wavelet coding compression and reconstruction error is coefficient quantization. Although the most widely used quantizers are uniform, the effectiveness of the quantization can be improved significantly by the following two methods (Gonzalez and Woods, 2002):
(i) Introducing an enlarged quantization interval around zero, called a dead zone. Though the increase of the quantization intervals will result in a larger number of truncated coefficients, it has a limitation. When the intervals attain a certain level, the number of the truncated coefficients will not be changed. In this case, the interval corresponds to the dead zone.
(ii) Adapting the size of the quantization interval from scale to scale. Since the gray-level distribution and the dynamic range of the image are related to the scale of the image, the suitable quantization intervals should be related to the scale of the image.

In either case, the selected quantization intervals must be transmitted to the decoder with the encoded image bit stream. The intervals themselves may be determined heuristically or computed automatically based on the image being compressed. For example, the median of detailed coefficients (with absolute values) can be used as a global threshold. Take a 512×512 Lena image, for instance. If it is transformed by the Daubechies 9/7 wavelet with a threshold of 8, then only 32,498 among 262,144 wavelet coefficients should be kept. In this case, the compression ratio is 8:1, and the PSNR of the reconstructed image is 39.14 dB.

6.5.6 Lifting-Based Wavelet Coding

A **lifting scheme** is a useful wavelet construction method (Sweldens, 1996). Wavelet transform based on a lifting scheme can realize an integer-to-integer transform in the current place. Therefore, if the transformed coefficients are directly coded, a lossless compression can be obtained. The decomposition process includes three steps.

6.5.6.1 Splitting
Splitting is used to decompose image data. Suppose that the original image is $f(x, y) = u_{j,k}(x, y)$. It can be decomposed into an even part $u_{j-1,k}(x, y)$ and an odd part $v_{j-1,k}(x, y)$. That is, the split operation can be written as (":=" denotes assignment)

$$S[u_{j,k}(x, y)] := [u_{j-1,k}(x, y), v_{j-1,k}(x, y)] \tag{6.48}$$

where $S[\cdot]$ is a splitting function, and

$$u_{j-1,k}(x, y) = u_{j,2k}(x, y) \tag{6.49}$$
$$v_{j-1,k}(x, y) = u_{j,2k1}(x, y) \tag{6.50}$$

6.5.6.2 Prediction

In the prediction step, the even part $u_{j-1,k}(x, y)$ is kept and used to predict the odd part $v_{j-1,k}(x, y)$, and then the odd part $v_{j-1,k}(x, y)$ is replaced by the differences (detailed coefficients) between the odd part $v_{j-1,k}(x, y)$ and the prediction value. This operation can be written as

$$v_{j-1,k}(x, y) := v_{j-1,k}(x, y) - P[u_{j-1,k}(x, y)] \tag{6.51}$$

where $P[\cdot]$ is a prediction function, and it performs an interpolation operation. The smaller the detailed coefficients are, the more accurate the prediction.

6.5.6.3 Updating

The goal of **updating** is to determine a better sub-image group $u_{j-1,k}(x, y)$, which can keep the characteristics, such as mean and energy, of the original image $u_{j,k}(x, y)$. This operation can be written as

$$Q[u_{j-1,k}(x, y)] = Q[u_{j,k}(x, y)] \tag{6.52}$$

In the updating step, an operator $U[\cdot]$ that acts on the detailed function $v_{j-1,k}(x, y)$ should be constructed, and the result of this action is added to the even part $u_{j-1,k}(x, y)$ to obtain the approximation image

$$u_{j-1,k}(x, y) := u_{j-1,k}(x, y) + U[v_{j-1,k}(x, y)] \tag{6.53}$$

Corresponding to the three operations for the decomposition process shown in eqs. (6.48), (6.51), and (6.52), the reconstruction process also includes three operations given by

$$u_{j-1,k}(x, y) := u_{j-1,k}(x, y) - U[v_{j-1,k}(x, y)] \tag{6.54}$$
$$v_{j-1,k}(x, y) := v_{j-1,k}(x, y) + P[u_{j-1,k}(x, y)] \tag{6.55}$$
$$u_{j,k}(x, y) := M[u_{j-1,k}(x, y), v_{j-1,k}(x, y)] \tag{6.56}$$

where $M[\cdot]$ is a merging function and $M[u_{j-1,k}(x, y), v_{j-1,k}(x, y)]$ performs the merging of the even part $u_{j-1,k}(x, y)$ and the odd part $v_{j-1,k}(x, y)$ to reconstruct the original image $u_{j,k}(x, y)$.

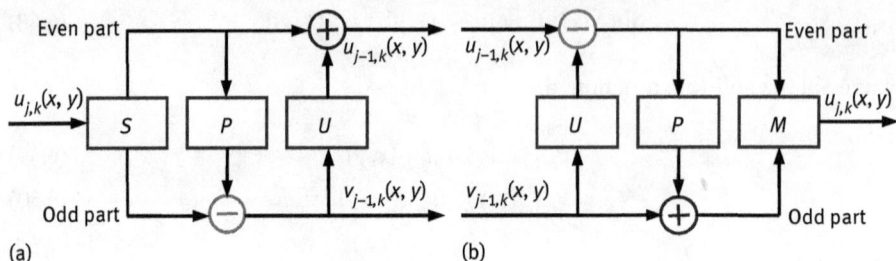

Figure 6.20: The decomposition and reconstruction of the lifting wavelet.

The whole procedures of decomposition and reconstruction with the lifting wavelet transforms are shown in Figures 6.20(a) and (b), respectively.

6.6 Problems and Questions

6-1* Suppose that the left and right images in Figure Problem 6-1 are the coding input and the output image, respectively. Compute the SNR_{ms}, SNR_{rms}, SNR, and PSNR of the output image.

$$\begin{array}{ccc} 2\ 4\ 8 & & 2\ 4\ 6 \\ 3\ 5\ 0 & & 3\ 4\ 0 \\ 3\ 7\ 8 & & 4\ 7\ 8 \end{array}$$

Figure Problem 6-1

6-2 (1) If a histogram equalized image with 2^n gray levels, could it still be compressed by a variable length coding procedure?

(2) Does such an image contain inter-pixel redundancies?

6-3 (1) For a source with three symbols, how many unique Huffman codes can be obtained?

(2) Construct these codes.

6-4 Using the source in Figure 6.3, construct Huffman codes, and obtain the code words and the average number of bits. Compare the results with that obtained by arithmetic coding.

6-5 It is assumed that the probabilities of symbols a, b, c are 0.6, 0.3, 0.1, respectively. The Huffman code obtained from such a sequence is 10111011001011. What is the most probable original symbol sequence?

6-6 It is assumed that the probabilities of symbols a, b, c, d, e are 0.2, 0.15, 0.25, 0.3, 0.1, respectively. Construct the arithmetic codes for the symbol sequence *bacddaecda*. If the accuracy is 32 bits, what is the maximum length of the symbol sequence that can be coded by the arithmetic coding method?

6-7 Using the gray-level transforms $T(r) = 0$, $r \in [0, 127]$ and $T(r) = 255$, $r \in [128, 255]$ can extract the seventh bit-plane in an 8-bit image. Propose a set of gray-level transforms capable of extracting all the individual bit-planes of an 8-bit image.

6-8 (1) Construct the entire 4-bit Gray code, and compare it with the 4-bit binary code.

   (2) Create a general procedure for converting a Gray-coded number to its binary equivalent and use it to decode 011010100111.

6-9 Suppose that one line in a bit-plane of an image is 11111011100001111000. Code it with 1-D run-length method and WBS method, respectively. Compare the bit numbers used in the two methods.

6-10 Change the predictor in eq. (6.24) to

$$f_n^* = \frac{f'_{n-1} + f_{n-1}}{2}$$

   Fill Table 6.8 and draw Figure 6.10 again.

6-11 Derive the Lloyd-Max decision and reconstruction levels for $L = 4$ and the uniform probability density function

$$p(s) = \begin{cases} 1/(2A) & -A \leq s \leq A \\ 0 & \text{otherwise} \end{cases}$$

6-12* Derive the Lloyd-Max decision and reconstruction levels for $L = 4$ and the probability density function:

$$p(s) = \begin{cases} s + k & -k \leq s \leq 0 \\ k - s & 0 \leq s \leq k \\ 0 & \text{otherwise} \end{cases}$$

6.7 Further Reading

1. Fundamentals

- To study the coding problem using information theory, some statistical models should be built (Gonzalez and Woods, 2002).

- A quality prediction model for judging JPEG format images can be found in Tsai and Zhang (2010).

- Other than using the objective and subjective criteria to judge the quality of coded images, application results can also be used as a criterion (Cosman *et al.*, 1994).

- In coding a color image, different color components can be used separately (Zhang *et al.*, 1997b).

- There must be some limits for a lossless coding. One method for estimating the bound can be found in Zhang *et al.* (1999).
- Besides lossless coding and lossy coding, the term "near-lossless coding" has also been used to indicate some constrained lossy coding, see (Zhang *et al.*, 2000) as an example.

2. **Variable-Length Coding**
   - Variable-length coding takes the entropy of an image as the coding bound. More information can be found in Salomon (2000), Gonzalez and Woods (2008), and Sonka *et al.* (2008).

3. **Bit-Plane Coding**
   - More introductions on other bit-plane decomposition methods and bit-plane coding methods can be found in Rabbani and Jones (1991).
   - Comparison of 1-D run-length coding and 2-D run-length coding can be found in Alleyrand (1992).

4. **Predictive Coding**
   - The principle of predictive coding can be used in video coding, which can be decomposed into intra-frame coding and inter-frame coding (Wang *et al.*, 2002).

5. **Transform Coding**
   - Fractal coding has a close relationship to transform coding (Barnsley, 1988).
   - Vector quantization (VQ) coding has many similarities with fractal coding (Gersho, 1982).
   - Recently, wavelet transform coding has attracted much attention (for an introduction to wavelets, see Chui (1992) and Goswami and Chan (1999)). Typical examples include embedded zero-tree wavelet (EWZ) coding and set partitioning in the hierarchical tree (SPIHT) coding, see (Said and Pearlman, 1996) and (Salomon, 2000).

7 Image Watermarking

The word "watermarking" has been in existence for several hundred years (watermarks on the banknotes and letterheads are typical examples). However, the term "digital watermark" appeared only in the late 1980s, while the real researches and applications of watermark on images started from the mid-1990s.

Image watermarking, a kind of image technology, can be divided into image space domain techniques and transform domain techniques, depending on its processing domain. Since many international standards for image coding and representation use discrete cosine transform (DCT) and discrete wavelet transform (DWT), image watermarking tasks are performed in DCT domain and DWT domain.

In general, image watermarking is a way of information hiding (embedding certain information invisibly in some kinds of carrier). Compared to other information hiding processes, watermarking is characterized by that the information contained in it is associated with the carrier.

The sections of this chapter are arranged as follows:

Section 7.1 introduces the principles and features of watermarks, including watermark embedding and detection, as well as the main characteristics of the watermarks. In addition, a variety of classification methods for watermarks are presented.

Section 7.2 describes the DCT domain image watermarking. The characteristics and principles of DCT domain watermarks are first discussed, and then the specific steps of a meaningless watermark algorithm and a meaningful watermark algorithm are detailed as examples.

Section 7.3 presents image watermarking in DWT domain. First, its features (in contrast to the DCT domain) are summarized. Then the focus is on the human visual system (HVS) and how to use its characteristics in watermarking algorithm. Finally, a watermarking algorithm in wavelet domain is introduced as an example.

Section 7.4 focuses on the watermark performance evaluation. A variety of image distortion measures are presented, some methods for watermark attacking are analyzed, and some benchmark measurements are discussed. In addition, the performance of the watermarking algorithms presented in the previous two sections are tested.

Section 7.5 discusses the watermark and related technology from a higher level. It introduces the relationships between watermarking and information hiding; it also describes an image hiding method based on iterative blending.

7.1 Principles and Characteristics

Digital **watermark** is a kind of digital signature. The watermark can be secretly embedded in the digital products (such as digital video, digital audio, digital picture,

DOI 10.1515/9783110524116-007

and electronic publications) to help identify the product owner, the right of use, the completeness of the contents, and so on. In the watermark, for example, a mark or code of the copyright owner and legal user can be found. These basic pieces of information can help establish a corresponding relationship between the digital products and their owners or users. The main uses of **image watermarking** include the following:

(1) Identification of the copyright: When the rights of the owners have been violated, the watermark can provide information to demonstrate that the owner is protected by the copyright of the image products (copyright).

(2) User identification: User's identity can be a legitimate record in the watermark, and be used to determine the source of the illegal copy.

(3) Confirm authenticity: The presence of the watermark can prove that the image has not been modified.

(4) Auto track: The watermark can be tracked by the system, so to know when and where the image has been used, which is important for levying royalty and for locating illegal users.

(6) Copy protection: The use of watermarks can regulate the use of the digital products, such as only playing video but not copying.

From the point of view of using watermark, the main operations of image watermarking are embedding and detection (extraction). From the point of view of watermarking performance, the different techniques have different characteristics, according to these characteristics, the different watermarks can be classified in a number of ways.

7.1.1 Embedding and Detection of Watermarks

Using the watermark to protect digital products requires two operations: one is to add a watermark for protection before using the product, commonly known as the **watermark embedding**; another is to extract embedded watermark from the product for verification or for indicating the copyright, commonly known as the **watermark detection.**

In order to effectively implement the processes of watermark embedding and detection, many models and techniques have been used (Cox *et al.*, 2002). The basic processes of watermark embedding and detection are introduced with the help of Figure 7.1. In Figure 7.1, the left part is for watermark embedding and the right part is for watermark detection. By embedding the watermark into the original image, the watermark embedded image can be obtained. If the relevant test is performed on the image to be detected, then whether or not a watermark is embedded in the image can be determined and the confidence of the judgment can also be obtained. In Figure 7.1, the detection of the watermark requires both the original image and the original watermark, but not all watermark detection systems are like this.

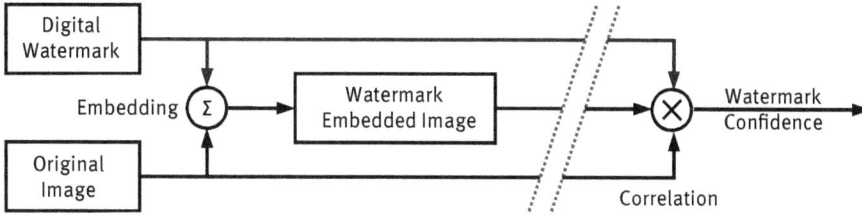

Figure 7.1: Diagram of watermark embedding and detection.

Let the original image be $f(x,y)$, watermark (is also an image) be $W(x,y)$, watermark embedded image be $g(x,y)$, then the process of watermark embedding can be described as:

$$g = E(f, W) \tag{7.1}$$

where $E(\cdot)$ represents the embedding function.

Once an image to be detected $h(x,y)$ is given, which may be the degraded version of the watermark embedded image $g(x,y)$ after transmission or processing, the possible authenticated watermark image is

$$w = D(f, h) \tag{7.2}$$

where $D(\cdot)$ represents the detection function.

Considering the original watermark and the potential watermark may be related by a correlation function $C(\bullet, \bullet)$, if a predetermined threshold value T is set, then

$$C(W, w) > T \tag{7.3}$$

signifies the existence of a watermark, otherwise no watermark exists. In practical situation, except the binary decision of exist or no exist, it is also possible to determine a confidence according to the correlation value, that is, proving a fuzzy decision.

From the point of view of signal processing, the watermark embedding process can be regarded as a process of superposition of a weak signal on some strong background; the watermark detection process can be regarded as a process of detecting weak signal from a noisy background. From the perspective of digital communication, the watermark embedding process can be considered as a process of using spread spectrum communication technology to transmit a narrow-band signal in a broadband channel.

7.1.2 Characteristics of Watermarks

The watermark will be embedded into the image according to the different purposes, so certain requirements need to be considered. Generally the most important characteristics of watermarks are the following.

7.1.2.1 Significance

Significance measures the no-perceptibility or (not easy) perceiving of watermark. For image watermark, this means the watermark is not visible. Significance includes two meanings: First, the watermark is not easily perceived by the user, the second is that a watermark added is not affecting the visual quality of the original product. From the perspective of human perception, image watermark embedding should not make the original image to have a perceptible distortion.

Significance is closely related to fidelity. The fidelity of a watermarked image can be measured by using the difference between the original image (without watermark) and the image to be judged. Considering the watermarked image in the transfer process may have been degraded, then the fidelity of a web image needs to be judged with comparison with the results of the original image after also the transmission.

Significance is a relative concept; the significance of watermark is dependent on the watermark itself and its contrast with the original image. As shown in Figure 7.2, the left is the original image; in the middle, a less transparent watermark have been added at a location with simple background, and the watermark is clearly seen; in the right, a more transparent watermark have been added at a location with complex background, and the watermark is hardly seen.

7.1.2.2 Robustness

Robustness refers to the ability of image watermark in resisting outside interference or distortion, under conditions still ensuring its integrity and detecting with required accuracy. In other words, the watermark in image should help to identify the works of ownership with extremely low error rate, so it is also known as reliability. Robustness of watermark is dependent on the quantity of embedded information and the strength of embedding (corresponding to the amount of embedded data).

In contrast to the requirements of watermark robustness, in the case to verify whether the original media have been changed or damaged, the fragile watermark

Figure 7.2: Significance is influenced by the position of watermark embedding and the background the watermark embedded on.

(also known as vulnerable watermark, that is, watermark with very limited robustness) should be used. Fragile watermarking can be used to detect whether the data protected by watermark have been modified. Fragile watermark is sensitive to the external processing (also known as sensitivity); it will change with the alternation of media. In this way, the alternation of the media can be determined according to the detected watermark, so as to determine whether the media has been modified or not.

7.1.3 Classification of Watermarks

According to the characteristics of technology, process, application, etc., there are many different ways to classify the watermark and the watermarking. Below are some of the typical examples.

7.1.3.1 Publicity Classification

From the publicity point of view, the watermark can be divided into the following four types.

(1) Private watermark: Its detection requires the original digital product. It is used to indicate the embedding location or to discriminate the watermark from other parts in images. Figure 7.1 has shown an example, in which the watermark detection requires the use of original watermark and/or original image.

(2) Half-private watermark: Its detection does not require the original digital products, but the information on whether there is watermark in these products.

(3) Public watermark: It is also called blind watermark. Its detection neither requires the original digital products nor the digital products having watermark embedded in. The watermark can be directly extracted from the obtained digital products. Compared to the blind watermark, the above two types of watermarks can be called non-blind watermarks.

(4) Asymmetric watermark: It is also called public key watermark. The watermark can be seen by all users but cannot be removed by any user. In this case, the key used in watermark embedding process is different from that in watermark detection process.

7.1.3.2 Perceptual Classification

From the perception (for image watermarking, this is also visibility) point of view, the watermark can be divided into the following two types.

Perceptual watermarks can be seen directly from the image, such as those marked on the web image for protect commercial utilization. Another example of visible watermark is the logo of TV channel. This kind of watermarks is not only to prove the ownership of the product but also to permit the appreciation of the product.

Non-perceptual watermark is also called invisible watermark, as using invisible ink technology to hide text in a book. This kind of watermark often indicates the identity of the original works; the fake should not be easy to go. It is mainly used for the detection of illegal copying as well as identification of products. The watermark cannot prevent illegal products from being illegally copied, but because the watermark exists in the product, it can be used as evidence in the court.

Image watermark is a kind of special watermark. For many digital products in general, adding visible signs in the product can be used to indicate the ownership. However, for the image, this may affect the visual quality and integrity. So the image watermark often refers to the invisible watermark, that is, the copyright owners know the existence of watermarks, but the watermarks are hidden for general users.

7.1.3.3 Meaning/Content Classification

From the point of view of meaning or content, the watermark can be divided into the following two types.

Meaningless watermark often use pseudorandom sequence (Gauss sequence, binary sequence, uniform distribution sequence, etc.) to express the presence or absence of the information. Pseudorandom sequence is formed by pseudorandom number, which is quite difficult to counterfeit, so it can guarantee the security of the watermark. For meaningless watermarks, hypothesis testing is often used.

No watermark present (n represents noise):

$$H0: g - f = n \tag{7.4}$$

Watermark present:

$$H1: g - f = w + n \tag{7.5}$$

Using pseudorandom sequence as a watermark, it can only provide two conclusions, "presence" and "no presence." The embedded information is equivalent to one bit. Pseudorandom sequence cannot represent concrete and specific information, so there are certain limitations in the applications.

Meaningful watermark can be text strings, seal, icons, images, and so on. It has specific and precise meaning, and is not easily forged or tampered with. In many identification or recognition applications, the watermark information includes the name, title, or logo of the owner. Such kind of watermark can provide more information than meaningless watermark, but it requires higher requirements for embedding and detection. The text watermark embedded in Figure 7.1 is an example of meaningful watermark.

7.1.3.4 Transform Domain Watermark

Although the image watermark can be embedded in space domain (image domain), most image watermark embedding is performed in the transform domain. The main advantages of transform domain watermarking include the following:
(1) The energy of the watermark signal can be widely distributed to all the pixels, which helps to ensure invisibility.
(2) Some characteristics of the human visual system may be more conveniently combined into the procedure, which helps to improve robustness.
(3) Transform domain watermarking is compatible with most international standards. It is also possible to perform watermarking directly in compressed domain (in this case, the watermark is known as bit stream watermark), thereby improving efficiency.

7.2 Image Watermarking in DCT Domain

In the DCT domain, the image may be decomposed into DC (direct current) coefficients and AC (alternating current) coefficients. From the perspective of robustness (in ensuring the watermark invisibility), the watermark should be embedded into the part of image where people feel the most important (Cox *et al.*, 2002). Therefore, the DC coefficient is more suitable than the AC coefficient for watermark embedding. From one side, comparing with AC coefficients, DC coefficients have much bigger absolute amplitude, so the DC coefficient has great capacity. From the other side, according to signal processing theory, the processes that watermarked image most likely have to encounter, such as compression, low-pass filtering, subsampling, and interpolation, affect more on AC coefficients than on DC coefficients. If the watermark is embedded into the AC coefficients and DC coefficients simultaneously, then the security will be enhanced by the embedding into AC coefficients and the amount of data will be increased by using DC coefficients.

The AC coefficients in the **DCT domain image watermarking** can be divided into two parts: high-frequency coefficients and low-frequency coefficients. The watermark embedded into the high-frequency coefficients could obtain good invisibility, but the robustness is poor. On the other hand, the watermark embedded into low-frequency coefficients could get better robustness, but the visual observation would be greatly influenced. In order to reconcile the contradiction between the two, the spread spectrum technology can be used (Kutter and Hartung, 2000).

7.2.1 Algorithm for Meaningless Watermark

The meaningless watermark is considered first. All algorithms for watermark consists of two steps: embedding and detection.

7.2.1.1 Watermark Embedding

A meaningless watermark scheme based on the combined utilization of DC coefficients and AC coefficients is introduced below (Zhang *et al.*, 2001b). The flowchart of **watermark embedding** algorithm is shown in Figure 7.3. Before the embedding of the watermark, the original image is preprocessed. The original image is first divided into small blocks, and all these small blocks are classified into two categories according to their texture: blocks with simple texture and blocks with complex texture. In addition, all blocks are transformed via DCT. Through the analysis of the DC coefficient (brightness) of each blocks, and combined with the results of the above classification, the image is finally divided into three categories:

(1) Blocks with low brightness and simple texture;
(2) Blocks with high brightness and complex texture; and
(3) Other blocks.

According to the principle of invisibility, the amount of watermark embedded in the first class should be smaller, while the amount of watermark embedded in the second class can be larger, and the amount of watermark embedded in the third class should be in the middle level.

It has been proved that the watermark with Gaussian random sequence has the best robustness (Cox *et al.*, 2002). Therefore, a random sequence of $\{g_m: 0, 1, \ldots, M-1\}$ obeying Gauss distribution $N = (0, 1)$ is generated as a watermark. The length of sequence M should be determined in considering both robustness and invisibility of watermark, and should also match the number of DCT coefficients. If four DCT coefficients are used for each block i, that is, $F_i(0, 0)$, $F_i(0, 1)$, $F_i(1, 0)$, and $F_i(1, 1)$, then M is equal to four times of blocks in image. According to the classification results of image blocks, each sample in this sequence is weighted by a suitable stretch factor and is embedded into DCT coefficients. Different embedding formulas for DC coefficients and

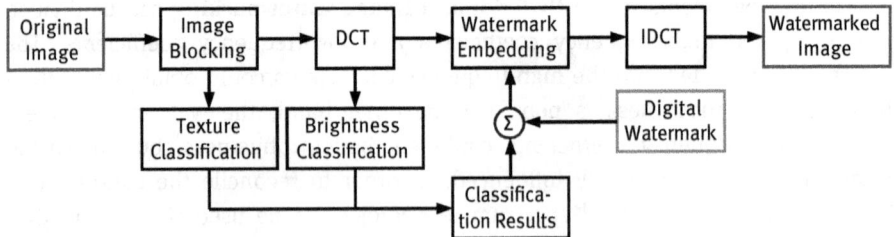

Figure 7.3: The flowchart of watermark embedding in DCT domain.

AC coefficients are used, a linear formula is used for AC coefficients and a nonlinear formula is used for DC coefficients. The combined formula is as follows:

$$F_i'(u, v) = \begin{cases} F_i(u, v) \times (1 + ag_m) & m = 4i & (u, v) = (0, 0) \\ F_i(u, v) + bg_m & m = 4i + 2u + v & (u, v) \in \{(0, 1), (1, 0), (1, 1)\} \\ F_i(u, v) & & \text{otherwise} \end{cases} \quad (7.6)$$

where a and b are stretch factors. According to Weber's law (the resolution of human visual system for relatively poor brightness), ag_m should be less than 0.02 in theory. In practice, for block with simple texture, taking $a = 0.005$; for block with complex texture, taking $a = 0.01$. The value of b is selected according to the previous classification: for category 1, taking $b = 3$; for category 2, taking $b = 9$; and for category 3, taking $b = 6$. Finally, performing IDCT for DCT domain image with adjusted coefficients to obtain the watermarked image.

Example 7.1 Invisibility of watermark
A set of sample results of watermark embedding obtained by the above method is given in Figure 7.4, wherein Figure 7.4(a) is a Lena image of 512×512, and Figure 7.4(b) is the watermarked image. Comparison of these two figures show that it is difficult to see from the visual feeling the presence of the watermark, even if the original image and the watermarked image are put together side by side. Figure 7.4(c) provides the difference between these two images, it is seen that almost no difference could be found. It is clear that the watermark embedded by this algorithm has good invisibility.

▣

7.2.1.2 Watermark Detection
The flowchart of **watermark detection** algorithm is shown in Figure 7.5. A hypothesis testing procedure is used for watermark detection. The difference between the original image and the image to be detected is calculated, the resulting image is divided into blocks, and further DCT is performed, from which the watermark sequence to be tested is extracted and compared with the original watermark via correlation checking, to identify whether it contains the watermark.

Figure 7.4: Comparison of original image and watermarked image.

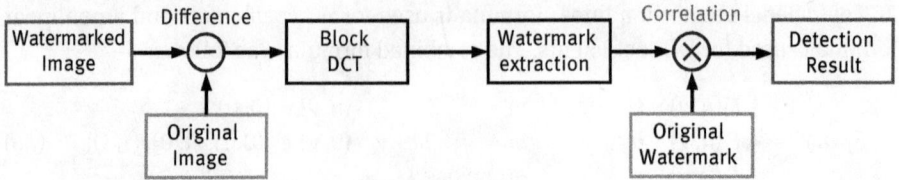

Figure 7.5: The flowchart of watermark detection in DCT domain.

Specific steps for watermark detection are as follows:

(1) Computing the difference image between original image $f(x, y)$ and the image to be tested $h(x, y)$ (supposing images are divided into I blocks with size 8×8):

$$e(x, y) = f(x, y) - h(x, y) = \bigcup_{i=0}^{I-1} e_i(x', y') \quad 0 \leq x', y' < 8 \tag{7.7}$$

(2) For each block in difference image, DCT is performed:

$$E_i(u', v') = \text{DCT}\{e_i(x', y')\} \quad 0 \leq x', y' < 8 \tag{7.8}$$

(3) Extracting the potential watermark sequence from block DCT coefficients:

$$w_i(u', v') = \{g_m, \ m = 4i + 2u' + v'\} = E_i(u', v') \tag{7.9}$$

(4) Computing the correlation between the original watermark and the potential watermark with the following function:

$$C(W, w) = \sum_{j=0}^{4I-1} (w_j g_j) \Bigg/ \sqrt{\sum_{j=0}^{4I-1} w_j^2} \tag{7.10}$$

For a given threshold T, if $C(W, w) > T$, the required watermark is detected, otherwise, no watermark. In selecting threshold, either false alarm or false detection is to be considered. For $N(0, 1)$ distribution, if the threshold was set as 5, then the probability of absolute value of watermark sequence bigger than 5 would be less than or equal to 10^{-5}.

7.2.2 Algorithm for Meaningful Watermark

The meaningless watermark contains only one bit of information, while meaningful watermark contains a number of bits of information.

7.2.2.1 Watermark Embedding

A meaningful watermark scheme, modified from the above meaningless watermark scheme, is introduced below (Zhang *et al.*, 2001b). Firstly, the symbol set (meaningful symbols) is constructed, whose length is L. Each symbol corresponds to a binary sequence of length M. For example, for each image block using the first four coefficients for embedding, then for a 256×256 image, the maximum number of bits could be embedded in images would be up to $4 \times (256 \times 256)/(8 \times 8) = 4,096$. If 32 bits are used to represent a symbol, then 4,096 bits can represent 128 symbols. Secondly, let the binary sequence of "0" and "1" obey Bernoulli distribution, so as to make the entire sequence with considerable randomness. Finally, when the number of symbols in the symbol set is less than the number of symbols that can be represented by the maximum number of bits embedded into image, the symbol will be repeatedly expanded. The length of the binary sequence will be expanded to an integer multiple of the length of symbols, and this expanded sequence will be added to the DCT coefficients of blocks.

Specific steps for **watermark embedding** are as follows:

(1) Dividing the original image into I blocks of 8×8, note each block as b_i, $i = 0, 1, \ldots, I - 1$:

$$f(x, y) = \bigcup_{i=0}^{I-1} b_i = \bigcup_{i=0}^{I-1} f_i(x', y') \quad 0 \le x', y' < 8 \tag{7.11}$$

(2) For each block, DCT is performed:

$$F_i(u', v') = \text{DCT}\{f_i(x', y')\} \quad 0 \le x', y' < 8 \tag{7.12}$$

(3) According to the length of symbol sequence to be embedded, select a suitable dimension M for the matching filter.

(4) Embedding the expanded sequence into DCT coefficients in blocks. Let $W = \{w_i | w_i = 0, 1\}$ be the expanded sequence corresponding to meaningful symbols, then the DCT coefficients with watermarks added are:

$$F'_i = \begin{cases} F_i + s & w_i = 1 \\ F_i - s & w_i = 0 \end{cases} \quad F_i \in D \tag{7.13}$$

where D represents the set of first four DCT coefficients, s represents the strength of watermark.

7.2.2.2 Watermark Detection

For meaningful watermark, the first three steps of **watermark detection** are the same as that of watermark detection for meaningless watermark. In the fourth step, suppose

for i-th detection, w_i^* is the strength of watermark, w_i^k is the output of k-th matching filter, thus their correlation is:

$$C_k(w^*, w^k) = \sum_{i=0}^{M-1} (w_i^* \cdot w_i^k) \bigg/ \sqrt{\sum_{i=0}^{M-1} (w_i^*)} \tag{7.14}$$

where M is the dimension of the matching filter. For a given j, $1 \le j \le L$, if

$$C_j(w^*, w^j) = \max\left[C_k(w^*, w^k)\right] \qquad 1 \le k \le L \tag{7.15}$$

Then, the symbol corresponding to j is the detected symbol.

7.3 Image Watermarking in DWT Domain

Compared with the DCT domain image watermarking, the advantage of **DWT domain image watermarking** benefits from the features of wavelet transform:

(1) Wavelet transform has multi-scale characteristic in space-frequency domain. The decomposition of the image can be performed continuously from low to high resolution. This helps to determine the distribution and location of the watermark so as to improve the robustness of the watermark and to ensure invisibility.

(2) DWT has fast algorithms for transforming the entire image. It also has good resilience to outside interference caused by filtering or compression processing. On the other side, performing DCT needs to divide image into blocks, which will produce mosaic phenomenon.

(3) The multi-resolution characteristics of DWT can better match with the human visual system (HVS), as seen below. This makes it easy to adjust the watermark embedding strength to accommodate the human visual system, so as to better balance the contradictions between robustness and invisibility of watermark.

7.3.1 Human Visual Properties

Through the observation and analysis of human vision phenomenon, and combining with the research results from visual physiology and psychology, it has been found that a variety of characteristics of HVS and masking properties (different sensitivity for different brightness/luminance ratio) are useful in the applications of image watermarking technology. Common characteristics and masking properties include the following:

(1) Luminance masking characteristic: The human eye is less sensitive to the additional noise in the high brightness region. This shows that the higher the image background brightness, the greater is HVS contrast threshold (CST), and the more the additional information can be embedded in.

(2)   Texture masking properties: If the image is divided into smooth area and texture area, then the sensitivity of HVS to the smooth region is much higher than
that to the texture region. In other words, the more complex of the texture in
background, the higher is HVS visibility threshold. As the existence of interference signal is harder to perceive in these regions, more information could be
embedded.

(3)   Frequency characteristic: The human eye has different sensitivity to the different
frequency components of the image. Experiments show that the human eye has a
low sensitivity to the high-frequency content and has a higher resolution for the
low-frequency region (corresponding to the smooth region).

(4)   Phase characteristics: The human eye has lower sensitivity to changes in phase
than changes in magnitude. For example, in watermarking based on discrete
Fourier transform, the watermarking signal is often embedded into the phase
part so as to improve the invisibility.

(5)   Directional characteristics: HVS is the most sensitive to the horizontal and vertical changes of the light intensity in the observation of the scene, while the
perception of the intensity change along the oblique direction is less sensitive.

With the help of the HVS characteristics and the visual masking properties, a visual
threshold (JND, for just notable difference) derived from HVS can be used to determine the maximum intensity in various parts of the image that can be tolerated by the
watermark signal (Branden and Farrell, 1996). This can thus avoid the destruction
of the visual quality of the image by embedding the watermark. In other words, the
human vision model is used to determine the modulation mask related to images.
This method can not only improve the invisibility of the watermark but also help to
improve the robustness of the watermark.

In the following, three visual masking properties of HVS are exploited, and the
visual threshold are determined accordingly (Barni *et al.*, 2001). Suppose here the
image undergoes an L-level wavelet decomposition (the number of sub-images is
$3L + 1$). In wavelet domain, the visual threshold based on human visual masking
properties can be expressed as $T(u, v, l, d)$, where u and v represent the location of
the wavelet coefficients, integer $l(0 \leq l \leq L)$ represents the wavelet decomposition level, $d \in \{LH, HL, HH\}$ represents the directions of high-frequency sub-band
images.

(1)   Human eye is not very sensitive to noise in various directions at different levels of
high-frequency sub-band images, and is also not very sensitive to the noise along
$45°$ direction in image (such as HH sub-band image). The sensitivity of different
sub-band images to the noise is inversely proportional to the masking factor. If
the masking factor for the noise in the l layer along the d direction is $M(l, d)$, then
$M(l, d)$ can be estimated by the following formula:

$$M(l, d) = M_l \times M_d \qquad (7.16)$$

where M_l and M_d consider the masking properties of different decomposition scales (corresponding to different layers) and different decomposition directions, respectively:

$$M_l = \begin{cases} 1 & l = 0 \\ 0.32 & l = 1 \\ 0.16 & l = 2 \\ 0.1 & l = 3 \end{cases} \tag{7.17}$$

$$M_d = \begin{cases} \sqrt{2} & d \in HH \\ 1 & \text{otherwise} \end{cases} \tag{7.18}$$

In eq. (7.17) and eq. (7.18), M_l takes larger values for high-frequency sub-band images, and small values for low-frequency sub-band images; M_d takes larger values for sub-band images along 45° direction, and small values for sub-band images along other directions. Large $M(l, d)$ value indicates that the sub-band sensitivity for noise is relatively low, so more watermarks can be superimposed on.

(2) Human eye has various sensitivities in different brightness regions. Usually, it is most sensitive to a medium gray region (the Weber ratio is held constant 0.02 for medium gray around a wide range) and is less sensitive in both lower and higher gray sides (with nonlinear decrease). In practice, such a nonlinear characteristic may be represented by quadratic curves. For example, the gray-level range of an 8-bit image can be divided into three parts, the dividing line between low part and middle part is at gray level 85 (threshold T_1 = 85), the dividing line between middle part and high part is at gray level 170 (threshold T_2 = 170). The normalized sensitivity curve is shown in Figure 7.6 by the convex lines, where the horizontal axis is the gray level axis. In the low gray level part, the sensitivity increases with the increase of the gray level as a second-order function. In the high gray level part, the sensitivity decreases with the increase of the gray level as a second-order function; in the middle of the gray level part, the sensitivity is kept constant.

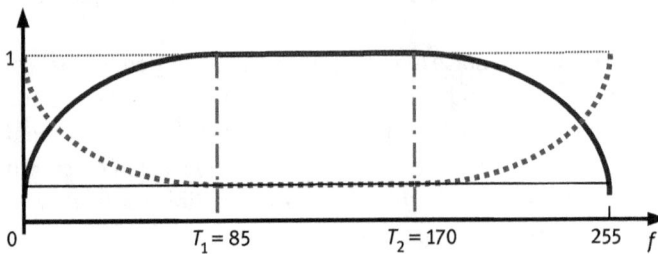

Figure 7.6: The normalized sensitivity curve.

In addition, the masking factor can be defined with the help of sensitivity curve. Dividing image into block, the average gray level of block is m, then the masking factor for noise at each point in this block is $B(u, v)$:

$$B(u, v) = \begin{cases} \dfrac{(0.2 - 0.02)[m - T_1]^2}{T_1^2} + 0.02 & m \le T_1 \\ 0.02 & T_1 < m \le T_2 \\ \dfrac{(0.2 - 0.02)[m - T_2]^2}{(255 - T_2)^2} + 0.02 & m > T_2 \end{cases} \qquad (7.19)$$

Thus obtained B curve is a concave curve, as shown in dotted line in Figure 7.6. From this curve, it is seen that at low and high gray levels, the noise sensitivity is relatively lower, so more watermark can be superimposed on.

(3) The human eye is more sensitive for smoothing image region and is less sensitive for textured image region. To discriminate these regions, the entropy for each region is calculated, small entropy indicates smooth region, while large entropy indicates textured region. The masking factor for a region can be determined by the entropy of this region. If the value of entropy in a block is H, normalizing the entropy of the block and multiplying it by the coefficient k to match other masking factors, a block texture masking factor can be obtained:

$$H(u, v) = k\frac{H - \min(H)}{\max(H) - \min(H)} \qquad (7.20)$$

In region with large texture masking factor, more watermark can be embedded.

Considering the above three factors in wavelet domain, the visual masking characteristic value is represented by the following formula (Barni *et al.*, 2001):

$$T(u, v, l, d) = M(l, d)B(u, v)H(u, v) \qquad (7.21)$$

The visual threshold provided by eq. (7.21) considers the human visual system sensitivity in different resolutions and different directions, as well as the masking effects for image blocks under different brightness contrast and different texture contents. According to the visual threshold, the intensity of the watermark embedding and the invisibility of the watermark can be controlled, so as to increase the strength of the embedded watermark and to enhance the robustness of the watermark.

7.3.2 Algorithm for Wavelet Watermark

In the following, the watermark embedding and watermark detection of an algorithm for wavelet watermark are described (Wang *et al.*, 2005).

7.3.2.1 Watermark Embedding

The basic flowchart of wavelet domain image watermarking method is shown in Figure 7.7, which shares some similarities with Figure 7.3.

Here Gauss distribution $N(0, 1)$ is used, the random sequence with length M is taken as watermark W, that is, $W = \{w_1, w_2, \ldots, w_M\}$.

The main steps of **watermark embedding** are as follows:

(1) Determining wavelet basis, performing L layer fast wavelet transform for original image $f(x, y)$, and obtaining a minimum frequency sub-band image and $3L$ different high-frequency sub-band images, respectively.

(2) According to eq. (7.21) to calculate the human eye visual threshold $T(u, v, l, d)$ in the high-frequency sub-band image. Then, according to thus obtained $T(u, v, l, d)$, the wavelet coefficients were arranged in descending order, and the top N wavelet coefficients were selected as the watermark insertion positions.

(3) Embedding watermark (that is, the watermark sequence is used to modulate the first N wavelet coefficients):

$$F'(u, v) = F(u, v) + qw_i \tag{7.22}$$

where $F(u, v)$ and $F'(u, v)$ are the (first N) wavelet coefficients of the original image and watermarked image, respectively; $q \in (0, 1]$ is the embedding strength; and w_i is the i-th components of the watermark sequence with length M. Here the embedding is carried out in DWT domain. In the process of the watermark embedding, the secrete key K is generated for extracting the watermark information, which is used to record the positions of the first N wavelet coefficients used in watermark embedding.

(4) Combining the low-frequency sub-band image with the high-frequency sub-band image in which watermark is embedded, performing fast inverse wavelet transform, so as to get the watermarked image $f'(x, y)$.

7.3.2.2 Watermark Detection

Watermark detection process can be seen as an approximation of the inverse of the watermark embedding.

(1) Selecting the wavelet basis used in the process of watermark embedding. Performing L layer wavelet decomposition for the original image $f(x, y)$ and the test

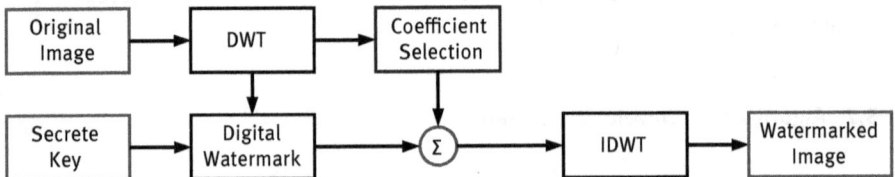

Figure 7.7: The flowchart of watermark embedding in DWT domain.

image $f''(x,y)$, respectively, to obtain the most low-frequency sub-band image and $3L$ high-frequency sub-band images. Note here that the test image $f''(x,y)$ may be different from the watermarked image $f'(x,y)$.

(2) According to the secret key K generated in the process of the watermark embedding, the important coefficient set $\{S_i, i = 1, 2, \ldots\}$ from the wavelet high-frequency sub-band channels in the original image $f(x,y)$ is obtained. Then, taking the address of these values as index, the corresponding coefficients from the wavelet high-frequency sub-band channels in the test image $f''(x,y)$ are selected as the important coefficient sets to be tested $\{S_i'', i = 1, 2, \ldots\}$. Comparing the values of S_i and S_i'', so as to extract the watermark information W''. When the difference between S_i and S_i'' is greater than a certain threshold, it can be declared there are watermark components w_i' at the corresponding position, its value can be set to 1, otherwise its value should be set to 0.

A quantitative evaluation of the similarity between the watermark sequence to be tested and the original watermark sequence can be made by using the normalized correlation coefficient C_N:

$$C_N(W, W'') = \frac{\sum_{i=1}^{L}(w_i - W_m)(w_i'' - W_m'')}{\sqrt{\sum_{i=1}^{L}(w_i - W_m)^2}\sqrt{\sum_{i=1}^{L}(w_i'' - W_m'')^2}} \tag{7.23}$$

where W and W'' are the original watermark sequence and the watermark sequence to be judged, W_m and W_m'' are the average values of W and W'', respectively. $C_N \in [-1, 1]$. If the value of C_N exceeds a certain threshold, it can be agreed that the W and W'' are related watermark sequences.

Example 7.2 Watermark distribution
A set of test results of watermark distribution and invisibility is given in Figure 7.8. The watermark used is a random sequence with Gaussian distribution $N(0, 1)$ and length $M = 1,000$. Figure 7.8(a) is an original Lena image, and Figure 7.8(b) is the watermarked image (PSNR = 38.52 dB). Figure 7.8(c) is the absolute difference image (it has been enhanced to increase contrast to make the difference more visible).

Two points can be remarked from Figure 7.8:
(1) From the visual perception point of view, it is hardly to see the difference between these two images (one before embedding and one after embedding), which shows that the algorithm has a good invisibility. In fact, the normalized correlation coefficient of these two images is 0.999, which means that the correlation between the two images is very high, which is consistent with the subjective feeling that these two images are quite similar.

Figure 7.8: Effect of watermark distribution and invisibility.

(2) From the difference image, it is seen that the watermark embedding intensity is larger in high texture region, low brightness region and high brightness region, but the watermark embedding intensity is relatively small in smooth region and the middle brightness region. In other words, the watermark embedding strength has the adaptive adjustment performance.

7.4 Performance Evaluation of Watermarking

The determination and evaluation of the performance of the watermarking is closely related to the characteristics and indexes of the watermark. Below several measures for the degree of distortion are first introduced, and some benchmark methods for performance evaluation are presented next. Finally, the watermark performance testing and evaluation are discussed by taking the performance verification of the watermarking algorithm described in the previous sections as examples.

7.4.1 Distortion Metrics

Watermark invisibility is a subjective indicator, more susceptible to the observer's experience and status, the test environment and conditions, and other factors. Some objective indicators (distortion metrics) can also be used. If the original image is $f(x, y)$, the watermark image is $g(x, y)$, the dimensions of these images are $N \times N$, then different distortion metrics can be defined. Commonly used **difference distortion metrics** include the following:.

(1) L^p Norm

$$D_{L^p} = \left\{ \frac{1}{N^2} \sum_{x=0}^{N-1} \sum_{y=0}^{N-1} |g(x, y) - f(x, y)|^p \right\}^{1/p} \tag{7.24}$$

When $p = 1$, it gives the mean absolute difference, when $p = 2$, it gives the root-mean-square error.

(2) Laplace Mean Square Error

$$D_{\text{lmse}} = \frac{\sum\limits_{x=0}^{N-1}\sum\limits_{y=0}^{N-1}\left[\nabla^2 g(x,y) - \nabla^2 f(x,y)\right]^2}{\sum\limits_{x=0}^{N-1}\sum\limits_{y=0}^{N-1}\left[\nabla^2 f(x,y)\right]^2} \tag{7.25}$$

Commonly used **correlation distortion metrics** include the following two.
(1) Normalized Cross-Correlation

$$C_{\text{ncc}} = \frac{\sum\limits_{x=0}^{N-1}\sum\limits_{y=0}^{N-1} g(x,y)f(x,y)}{\sum\limits_{x=0}^{N-1}\sum\limits_{y=0}^{N-1} f^2(x,y)} \tag{7.26}$$

(2) Correlation Quality

$$C_{\text{cq}} = \frac{\sum\limits_{x=0}^{N-1}\sum\limits_{y=0}^{N-1} g(x,y)f(x,y)}{\sum\limits_{x=0}^{N-1}\sum\limits_{y=0}^{N-1} f(x,y)} \tag{7.27}$$

7.4.2 Benchmarking and Attack

In addition to the design of watermarking technology, the development and utilization of the evaluation and benchmarking of the watermarking technology have also been a lot of research.

7.4.2.1 Benchmarking Methods
There are many ways to determine the performance of the watermarking with baseline measurement (**benchmarking**). In general, the robustness of the watermarking is related to the visibility and payload of the watermarking. For fair evaluation of different watermarking methods, it is possible to first determine certain image data, in which watermark is embedded as much as possible while it does not very much affect the visual quality. Then, the embedded watermark data are processed or attacked, so the performance of the watermarking methods can be estimated by measuring the error ratio. It is seen that the benchmarking for measuring watermarking performance will be related to the selection of the payload, the visual quality measurement process as well the procedure of attack. Here are two typical benchmarking methods.

Table 7.1: Processing methods and associated parameters for judging robustness.

#	Process operation	Parameter
1	JPEG compression	Quality factor
2	Blurring	Mask size
3	Noise	Noise levels
4	Gamma correction	Gamma index
5	Pixel exchange	Mask size
6	Mosaic (filtering)	Mask size
7	Median filtering	Mask size
8	Histogram equalization	

Robustness Benchmarking In this benchmarking (Fridrich and Miroslav, 1999), the payload is fixed at 1 bit or 60 bits; visual quality measure is adopted from spatial mask model (Girod, 1989). This model is based on the human visual system, and it accurately describes the visual distortion/degradation/artifact (artefact) produced in edge regions and smooth regions. The strength of the watermark is adjusted so that only less than 1% of the pixels can be seen when these pixels have been changed, based on the model above. The selected processing methods and related parameters are listed in Table 7.1; the visual distortion ratio is a function of the relevant process parameters.

Perception Benchmarking In this benchmarking (Kutter and Petitcolas, 1999), the effective payload is fixed to 80 bits; the visual quality metric used is the distortion metric (Branden and Farrell, 1996). This distortion metric considers the contrast sensitivity and the mask characteristic of the human vision system and counts the points over the visual threshold (JND).

7.4.2.2 Outside Interferences

There are a variety of outside interferences for images. They can be divided into two categories from the perspective of the robustness of the watermark. The first category consists of the conventional means of image processing (not specifically for watermark), such as sampling, quantization, D/A and A/D conversion, scanning, low-pass filtering, geometric correction, lossy compression, and printing. The ability of image watermarking against these outside interferences is more commonly called the robustness. The second category refers to malicious attack mode (specifically for watermarking), such as illegal detection and decoding of the watermark, resampling, cropping, special displacement, and the scale changes. The ability of watermarking to withstand these outside attacks is commonly measured by anti-attack aptitude.

The attacks to watermark refers to various unauthorized operations, they are often divided into three types:

(1) Detection: For example, a user of watermark product tries to detect a watermark that only the owner has the right to do, which is also called **passive attacks**.

(2) Embedding: For example, a user of watermark product tries to embed a watermark that only the owner has the right to do, which is also called **forgery attack**.

(3) Deletion: For example, a user of watermark product tries to delete a watermark that only the owner has the right to do, which is also called *removing attack* that can be further divided into **eliminating attack** and **masking attack**.

Several types of the above attacks may also be used in combination. For example, first to remove existing watermark from products and then to re-embed other watermarks needed, which is also called **changing attack**.

The so-called watermark attack analysis is to design methods to attack the existing watermarking system, in order to test its robustness. The purpose of the attack is to make the corresponding watermarking system from correctly restoring watermark or make the detecting tool from detecting the presence of the watermark. By analyzing the weaknesses and vulnerability of system, it is possible to improve the design of watermarking systems.

There is a simulation tool called StirMark, which is a watermark attack software. People can use it to evaluate the different anti-attack capability of watermarking algorithms, by inspecting whether the watermark detector can extract or detect watermarks from watermark carrier that has suffered attacks. StirMark can simulate many kinds of processing operations and various means of attack, such as geometric distortion (stretching, shear, rotate, etc.), nonlinear A/D and D/A conversion, print out, scan, and resampling attacks. In addition, StirMark can also combine a variety of processing operations and means of attack to form a new kind of attack.

7.4.3 Examples of Watermark Performance Test

In the following, some performance test processes and results for the watermarking algorithms presented in the above two sections are given.

7.4.3.1 Meaningless Watermarking in DCT Domain

Two tests using Lena image are carried out.

Robustness Test Several common image processing operations and interferences are tested with the watermarking algorithm presented in Section 7.2.1. Figure 7.9 shows several results obtained for watermarked Lena image. Figure 7.9(a) is the result of mean filtering with mask of size 5×5 (PSNR = 21.5 dB). Figure 7.9(b) is the result of

Figure 7.9: Verify examples for image processing operations on watermark.

2: 1 subsampling in both the horizontal and vertical directions (PSNR = 20.8 dB). Figure 7.9(c) is the result of compression by retaining only the first four DCT coefficients (PSNR = 19.4 dB). Figure 7.9(d) is the result of adding white Gaussian noise (PSNR = 11.9 dB, the effect of noise is quite obvious). Under these four situations, the images all have exceedingly distortion, but the watermark can still be accurately detected out.

Uniqueness Test   Given a random sequence as the watermark, other sequences generated by the same probability distribution could also be taken as watermarks. According to this idea, 10,000 random sequences are generated from a Gaussian distribution $N(0, 1)$, taking one of them as a (real) watermark and all others as fake watermarks for comparison. The test results of mean filtering, subsampling, and compression processing are shown in Table 7.2. Since the differences between the real watermark and the results obtained with fake watermark are very significant (correlation values are quite distinct), the real watermark and the fake watermark can easily be distinguished.

7.4.3.2 Meaningful Watermarking in DCT Domain
In addition to the use of Lena image, the performance test of the meaningful watermarking algorithm presented in Section 7.2.2 also uses two other images: the Flower and the Person, respectively, as shown in Figures 7.10(a) and (b).

Three tests are carried out.

Robustness Test against Mean Filtering   A sequence of eight symbols is embedded in each test image. Masks of 3×3, 5×5, and 7×7 are used, respectively. Figure 7.11 gives a few of the filtering results. Figures 7.11(a) and (b) are for Lena image with masks of

Table 7.2: Test results on the uniqueness of watermark.

Image processing operation	Original image	Mean filtering	Subsampling	Compression
Correlation with real watermark	114.5	13.8	23.6	12.4
Correlation with fake watermarks (Max)	3.82	4.59	3.58	3.98

Figure 7.10: Two other test images.

Figure 7.11: Test results after mean filtering.

5×5 and 7×7, respectively. Figures 7.11(c) and (d) are for Flower image with masks of 5×5 and 7×7, respectively.

Detection of meaningful watermark needs not only to extract all symbols but also to detect correctly the position of each symbol. Table 7.3 gives the corresponding test results, the number of correct symbols (# Correct) here refers to both the correct symbols and the correct position. In each case, the peak signal-to-noise ratio (PSNR) of watermarked images after low-pass filtering is also provided to give an indication of the quality of the image.

It is seen from Figure 7.11 and Table 7.3 that with the increase in the size of the mask, the image becomes more blurred, and the ability of watermark to resist mean filtering drops, too. Since the details of Lena image and Flower image are less than the details of Person image, so after filtering with 5×5 mask, all watermark symbols

Table 7.3: Robustness test results of mean filtering.

Image	3 × 3 Mask		5 × 5 Mask		7 × 7 Mask	
	PSNR/dB	# Correct	PSNR/dB	# Correct	PSNR/dB	# Correct
Lena	25.6	8	21.5	8	19.4	3
Flower	31.4	8	25.6	8	22.5	3
Person	19.3	8	12.4	4	9.4	1

can still be detected, but at this time the watermark symbols embedded in the Person image could not be correctly detected.

Robustness Test against Subsampling   A sequence of eight symbols is embedded in each test image. Three kinds of subsampling rate (both in the horizontal and vertical directions) are considered here: 1:2 subsampling, 1:4 subsampling, and 1:8 subsampling. Figure 7.12 gives the results of the several subsampling process, in which Figures 7.12(a) and (b) are the partial view of Lena image after 1:2 and 1:4 subsampling, respectively; Figures 7.12(c) and (d) are the partial view of Person image after 1:2 and 1:4 subsampling, respectively. Comparing two groups of images, the subsampling has more influence on Person image than on Lena image.

Table 7.4 lists the corresponding test results. In various situations, the peak signal-to-noise ratio (PSNR) of the watermarked images after subsample is also listed in order to give an indication of the quality of the image. Compared to the cases of mean filtering, the subsampling leads to a more serious distortion of the image.

Robustness Test against JPEG Compression   In this experiment, the sequence size of embedded symbols varies from 1 to 128, for each test images. Table 7.5 gives the corresponding test results. Values given in the table is the lowest value of an image PSNR at which the entire sequence of watermark symbols can be correctly detected. As can be seen from Table 7.5, the length of sequence of symbols embedded is related with the image PSNR. Since the dimension of the matched filter is automatically adjusted,

Figure 7.12: Several result images after subsampling.

Table 7.4: Robustness test results of sub-sampling.

Image	1 : 2 Subsampling		1 : 4 Subsampling		1 : 8 Subsampling	
	PSNR/dB	# Correct	PSNR/dB	# Correct	PSNR/dB	# Correct
Lena	20.8	8	15.5	4	12.0	1
Flower	24.2	8	17.2	5	13.5	1
Person	12.8	8	6.6	0	4.6	0

Table 7.5: Robustness test results of JPEG compression (dB).

Length Image	1	2	4	8	16	32	64	128
Lena	23.4	25.4	26.5	27.2	28.3	30.5	34.5	36.5
Flower	15.4	20.4	20.6	20.6	21.6	22.2	35.3	38.3
Person	23.3	24.3	25.8	26.7	26.8	26.8	26.9	38.4

Table 7.6: The resistance of watermark to some image processing and attacks.

Processing/Attacks	PSNR/dB	Normalized correlation coefficient C_N
Mean filtering	19.35	0.908
Gaussian noise	16.78	0.537
JPEG compression (compression ratio 37:1)	27.15	0.299
Median filtering	32.00	0.689
Sharpening	34.19	0.969
2×2 Mosaic	29.62	0.531

so for the sequence of symbols of different lengths, the noise level that can be tolerated is also different.

7.4.3.3 Watermarking in DWT Domain

The robustness test for the wavelet domain watermarking algorithm described in Section 7.3 still uses Lena image. The resistance of watermark embedded to some image processing processes and attacks is shown in Table 7.6, if the decision threshold is set to 0.2, all the symbols of watermark can be detected correctly according to the normalized correlation coefficients.

7.5 Information Hiding

Information hiding is a relatively broad concept, generally refers to embed some specific information intentionally hidden into a carrier, in order to achieve the purpose of a secret.

7.5.1 Watermarking and Information Hiding

Information hiding can be secret or non-secret, depending on whether the existence of information is confidential or not confidential. In addition, information hiding can be the type of watermark or the type of non-watermark, depending on whether these specific information are correlated with or not correlated with the carrier.

Table 7.7: The classification of information hiding technology.

	Correlated with carrier	No correlation with the carrier
Hiding existence of information	(1) Secret watermark	(3) Secret communication
Knowing existence of information	(2) Non-secret watermark	(4) Secret embedding communications

According to the above discussion, the **information hiding** technology can be divided into four categories, as shown in Table 7.7 (Cox *et al.*, 2002):

7.5.2 Images Hiding Based on Iterative Blending

Image hiding can be viewed as a special kind of steganography. It intends to hide an image into the carrier image for transmission. In actual utilization, the carrier image is generally a common image that can be passed without public suspicion. In the following, a way to hide image is presented.

7.5.2.1 Image Blending

Let carrier image be $f(x, y)$, the hidden image be $s(x, y)$. Suppose α is a real number satisfying $0 \leq \alpha \leq 1$, then the image

$$b(x, y) = \alpha f(x, y) + (1 - \alpha)s(x, y) \tag{7.28}$$

is the result of blending images $f(x, y)$ and $s(x, y)$ with parameter α, which is called a trivial blending when α is 0 or 1.

A blending example of two images is shown in Figure 7.13, in which Figure 7.13(a) is the carrier image (Lena image) and Figure 7.13(b) is the hidden image (Girl image). Taking $\alpha = 0.5$, the blending image in as Figure 7.13(c).

Figure 7.13: An example of image blending.

From the perspective of camouflage, the image obtained by blending should not be different in the visual sense with carrier image. According to the definition of the **image blending**, when the parameter α is close to 1, the image $b(x, y)$ will be close to $f(x, y)$; when the parameter α is close to 0, the image $b(x, y)$ will be close to $s(x, y)$. This allows the utilization of human visual characteristics for better hiding one image into another image. Hidden images can be restored by the following formula:

$$s(x, y) = \frac{b(x, y) - \alpha f(x, y)}{1 - \alpha} \tag{7.29}$$

For digital images, some errors in rounding will be produced during the calculation and recovery process, and this induces the drop of quality of restored image. This error is dependent on the two images themselves and the blending parameter. When two images were given, this error is only the function of parameter α. This error can be measured by the root-mean-square error between two images. In Figure 7.14, the function curves of rounding error versus the parameter α (left for blended image, right for restored image) are given, where Figures 7.13(a) and (b) are taken as carrier image and hidden image, respectively (Zhang *et al.*, 2003).

It is seen from Figure 7.14 that the more the parameter α approaches 1, the better the effect of image hiding, but the worse of the quality of restored image. On the contrary, if the better quality of restored image is required, then the parameter α could not approach 1; however, the effect of image hiding would not be good. Therefore, there must be a best blending parameter value that can make the sum of errors from blended image and from restored image to be a minimum, this is shown by the valley in the curve of Figure 7.15.

In summary, the general principles of image hiding are selecting first the carrier image that resemble the image to be hidden as close as possible, then selecting the blending parameter as small as possible in the permitted visual range. In this way, the quality of the restored image could be ensured.

7.5.2.2 Iterative Blending with Single Image
Extending the above procedure, it is possible to obtain the iterative blending of images by using several blending parameters with a number of times.

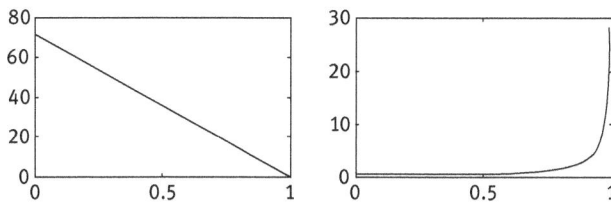

Figure 7.14: Relationship between the qualities of blended image and restored image with different blending parameters.

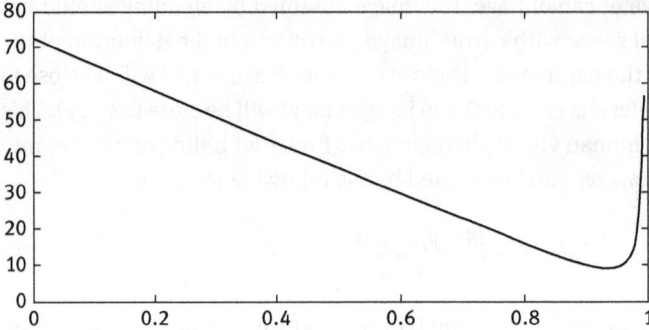

Figure 7.15: The curve of best blending.

Suppose $\{a_i | 0 \le a_i \le 1, i = 1, 2, \ldots, N\}$ are N given real numbers, blending image $f(x, y)$ and image $s(x, y)$ with parameter a_1 gives $b_1(x, y) = a_1 f(x, y) + (1 - a_1)s(x, y)$, blending image $f(x, y)$ and image $b_1(x, y)$ with parameter a_2 gives $b_2(x, y) = a_2 f(x, y) + (1 - a_2)b_1(x, y), \ldots$, continuing this procedure can obtain $b_N(x, y) = a_N f(x, y) + (1 - a_N)b_{N-1}(x, y)$, and image $b_N(x, y)$ can be called N-fold iterative blend image with respect to $\{a_i, i = 1, \ldots, N\}$. It can be proved that in no-trivial cases, image $b_N(x, y)$ will monotonically be converged to the carrier image $f(x, y)$:

$$\lim_{N \to \infty} b_N(x, y) = f(x, y) \tag{7.30}$$

Figure 7.16 presents some examples of using the above iterative algorithm for image hiding and restoration, in which the images of Figures 7.13(a) and (b) are taken as carrier image and hiding image, respectively. In Figure 7.16, the up line images are blending results with 1, 2, and 3 iterations (the blending parameters are 0.8, 0.7, and 0.6), respectively; the bottom line images are the hiding images restored from corresponding blend images. The related parameters and error data are listed in Table 7.8, with RMSE stands for root-mean-square error.

7.5.2.3 Iterative Blending with Multiple Images

The above algorithm of image blending and algorithm of **iterative blending with single image** embedding both a secret image in only one carrier image. If the attacker does intercept the carrier image and blend image, and does have some doubt, then the attacker would restore the secret image with the help of the original carrier image by subtraction. The security of such a hiding system is completely dependent on only one carrier image, so it is relatively fragile. In order to solve this problem, the idea of image blending can be extended to use multiple blending parameters and also multiple carrier images to hide a single secret image, which is called the **iterative blending with multiple images**.

Figure 7.16: Experiment results of iterative blending with one carrier image.

Table 7.8: Parameters and error data of iterative blending with single image.

Blending parameter	0.8	0.7	0.6
PSNR of blending image/dB	24.9614	35.4190	43.3778
RMSE of blending image	14.4036	4.3211	1.7284
PSNR of restored image/dB	45.1228	34.5148	26.3956
RMSE of restored image	1.4138	4.7951	12.2112

Let $f_i(x, y), i = 1, 2, \ldots, N$ be a group of carrier images, $s(x, y)$ be a secret image, $\{\alpha_i | 0 \leq \alpha_i \leq 1, i = 1, 2, \ldots, N\}$ be N given real numbers. Blending image $f_1(x, y)$ and image $s(x, y)$ with the parameter α_1 gives $b_1(x, y) = \alpha_1 f_1(x, y) + (1 - \alpha_1)s(x, y)$, Blending image $f_2(x, y)$ and image $b_1(x, y)$ with the parameter α_2 gives $b_2(x, y) = \alpha_2 f_2(x, y) + (1 - \alpha_2)b_1(x, y)$, continuing this procedure can obtain $b_N(x, y) = \alpha_N f_N(x, y) + (1 - \alpha_N)b_{N-1}(x, y)$. The image $b_N(x, y)$ is called N-fold iterative blend image with respect to parameters $\{\alpha_i, i = 1, \ldots, N\}$ and images $\{f_i(x, y), i = 1, \ldots, N\}$.

According to the definition of iterative blending with multiple images, it is possible to obtain an image hiding scheme, which blend a secret image into multiple carrier images iteratively with the help of masking properties of human visual system. To restore such a secret image, it is required to use N blend images and N blend parameters, and to know the blend order of these images. Therefore, this scheme of iterative blending with multiple images is a quite safe one.

Figure 7.17: An example of iterative blending with multiple carrier images.

One example of iterative blending with multiple images is shown in Figure 7.17. In the hiding process, the image Couple in Figure 7.17(c) is hided in the image Girl in Figure 7.17(b) with blend parameter $\alpha_2 = 0.9$, the result image is further hided in the image Lena in Figure 7.17(a) with blend parameter $\alpha_1 = 0.85$. In this example, Figure 7.17(a) is the open image, Figure 7.17(b) is the image with intermediate result, and Figure 7.17(c) is the hiding image.

7.6 Problems and Questions

7-1 What are the differences between the problems encountered in the embedding and extraction of watermarks?

7-2* What are the other properties/features of the watermark, except those that have been introduced in the text? Give a few examples.

7-3 If dividing the brightness of the image block into three levels: low brightness, medium brightness, high brightness; and dividing the texture of the image block also into three types: simple texture, medium texture, complex texture, then there are nine combinations of them. Please make an analysis on the influence of watermark embedding in these nine cases from the perspective of invisibility.

7-4* If you want to embed "TSINGHUA UNIVERSITY" four times in a 256 by 256 image by using the method described in Section 7.2.2, how many coefficients from each image block should be used at least?

7-5 It is required to provide one example each from the everyday life for the three types of human visual masking features described in Section 7.3.1.

7-6 Please make an analysis on the characteristics (such as advantages, disadvantages) of each distortion measure described in Section 7.4.1. What are the suitable application areas for each of these measures?

7-7 Select an image, and make lossy JPEG compression on it. With the original image and compressed images, calculate the distortion measures introduced in

Section 7.4.1. Draw the curve of each distortion measure value as the function of the compression rate.

7-8   What is the difference between a malicious attack and a conventional image processing operation on a watermark? Give some examples for each of them.

7-9   Collect and create a list of malicious attacks on watermarks to discuss how they affect the presence and detection of watermarks?

7-10   After adding a certain amount of watermark information to an image, the distortion of the watermarked image can be measured by using the distortion measures described in Section 7.4.1. Compare these measures according to the (relative) values of these distortion measure results to rank them. Compare further these measures with the subjective sensation.

7-11   Many watermarking applications can also be accomplished by using other techniques, so what are the advantages of watermarking over other technologies?

7-12   What is the relationship between image watermarking and image coding? What are their similarity and difference?

7.7 Further Reading

1. **Principles and Characteristics**
   - A comprehensive description of watermarking techniques is also available in Cox *et al.* (2002).
   - Some discussions on the future of watermarking can be found in Barni *et al.* (2003a, 2003b).

2. **Image Watermarking in DCT Domain**
   - More introduction to various algorithms can be found in Shih (2013).

3. **Image Watermarking in DWT Domain**
   - One of the features of DWT domain image watermarking is the easy integration of human visual characteristics (Barni *et al.*, 2001).

4. **Performance Evaluation of Watermarking**
   - There are publicly available tools to test the robustness of image watermarking techniques. For example, there is a tool available for JPEG format images called Unzign (1997).

5. **Information Hiding**
   - Although there are certain differences between steganography and watermarking, their components complement each other more than they compete (Kutter and Hartung, 2000).

8 Color Image Processing

Perception of color is the inherent ability of the human visual system (HVS). Although color vision is subjective, and it cannot be fully explained with the results from theoretical research and practice, much knowledge about the physical nature of color has already been gathered. For example, a lot of attributes of color vision, such as wavelength resolution, color saturation, its relationship with brightness, and the rules of color mixing, have been clearly known. On the other hand, with the progress of technology in recent years, many devices for color image acquisition and processing are widely spread and used in a large number of applications. Therefore, the color image becomes also the subject of image processing.

Compared to black-and-white images, color images contain more information. In order to effectively represent and manipulate color information, it is necessary to establish the corresponding color representation models, and to study the corresponding color image processing technology.

Color image processing technologies can be divided into two categories. First, because people's ability to distinguish different colors and sensitivity to color would be stronger than gray scale, converting a gray-level image to a color image should increase the efficiency in observing the image content. Such an image processing technique is often referred to as pseudo-color processing technology. Second, the color image can also be directly subjected to various treatments to achieve the desired effects. The image processing technology related to pure color images is the true color processing technology.

The sections of this chapter are arranged as follows:

Section 8.1 introduces the principles of color vision and expresses the methods for color description. It includes the fundamentals of color vision, the three primary colors and color matching, as well as an explanation for color chromaticity diagram.

Section 8.2 presents several basic and common color models, including the device-oriented RGB and CMY models; I_1, I_2, I_3 model; a normalized model; a color TV model; a video model; and some color models suitable for visual perception (such as HSI, HSV, HSB models, and $L^*a^*b^*$ model). The relationships among these models are also discussed.

Section 8.3 describes several basic pseudo-color image enhancement techniques, including both the spatial domain and frequency domain techniques.

Section 8.4 discusses the strategy for true color image processing, and introduces the transform enhancing methods by separately treating each color component and the filter enhancing methods by considering all color components simultaneously (including both linear and nonlinear methods).

DOI 10.1515/9783110524116-008

8.1 Color Vision and Chromaticity Diagram

To carry out the color image processing, it is required first to understand the sensing principle and expression of color.

8.1.1 Color Vision Fundamentals

Color vision is related to the physical essence of color and human vision system. The essence of color was systematically studied and discovered first by Newton. In the 17th century, Newton found, by using prism to study white light refraction, that the white light can be decomposed into a series of continuous spectrum ranged from purple to red. It is proved that the white light is made up of different color lights (and these colors cannot be further decomposed). In other words, white light is mixed and composed of different color lights. These different color lights are electromagnetic waves of different frequencies. Not only can human vision perceive the light stimulus, but it can also perceive the electromagnetic wave of different frequencies as different colors. In the physical world, the power distribution of radiation energy is objective, but color exists only in the human eye and brain. Newton once said: "Indeed rays, properly expressed, are not colored" (Poynton, 1996).

When color refers to different hues or pigments, noncolor refers to white and black as well as various grays that are different degree of shades between white and black. The surface that can absorb all wavelengths of light (no selective reflection spectrum) looks gray. If it reflects more light it looks as light gray, while it reflects less light it looks dark gray. In general, if the reflected light is less than 4% of the incident light, the surface looks black, while if more than 80–90% of the incident light is reflected, the surface looks white.

The physical basis of color vision is that the human retina has three types of color photoreceptor cone cells, which respond to incident radiation with somewhat different spectral response curves. In other words, these cone cells are sensitive to different wavelengths of radiation. There is a fourth type of photoreceptor called rod cell in the retina. Rods are effective only at extremely low light levels. Because there is only one type of rod cell, "night vision" cannot perceive color.

Physiological basis of color vision and visual perception is related to chemical processes and neural processes in the brain nervous system. Overall, the human color vision has a series of elements, corresponding to a complex process. First, the creation of color vision needs a light source. The light from source is transmitted through the reflection or transmission mode to attend the eye, and to cause nerve signals to be received by retinal cells, the human brain finally explains this to produce color vision.

The human feeling of the object color depends on the characteristics of the reflected light. If the reflection of object has a more or less balanced spectrum, the human perception of object is white or gray. If the reflection of object has more spectral reflectance in certain frequencies, the human perception of object would be corresponding to related colors.

8.1.2 Three Primary Colors and Color Matching

Three primary colors are the basic units of color production, which can produce a number of color combinations. The combination colors would be different than the colors in the combination. This is process is also called **color matching**.

8.1.2.1 Three Primary Colors
The three colors corresponding to cone cells with different retinal color feelings are called three primary colors (the mixing of either two colors does not produce a third color). The **three primary colors** are red (R), green (G), and blue (B). These colors correspond to the three kinds of cone cells, which have different wavelength response curves to external radiation, as shown in Figure 8.1.

It is noted that the three response curves in Figure 8.1 are distributions with wider ranges and there are certain overlaps. In other words, a particular wavelength of light can stimulate two to three kinds of cells to make them excite. Therefore, even if the incident light has a single wavelength, the reaction of the human visual system is not simple. Human's color vision is the result of a combined response to different types of cells. In order to establish a standard, the International Commission on Illumination (CIE), as early as in 1931, has specified the wavelength of the three primary colors, red, green, and blue (R, G, B) to be 700 nm, 546.1 nm, and 435.8 nm.

8.1.2.2 Color Matching
When mixing the red, green, and blue colors, the obtained combination color C may be regarded as a weighted sum of the strength proportion of the three color

$$C \equiv rR + gG + bB \tag{8.1}$$

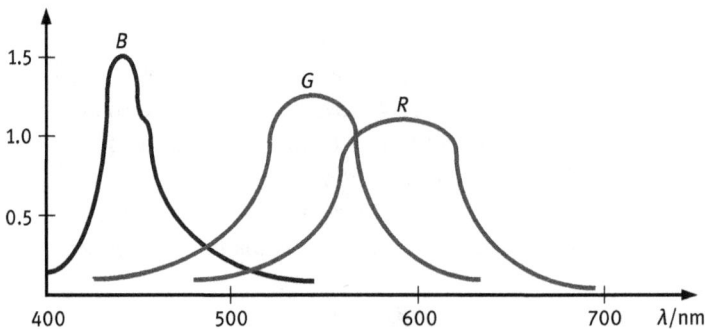

Figure 8.1: Wavelength response curves of three cone cells.

where $\equiv$ represents matching, r, g, b represents ratio coefficients for R, G, B, and there is $r + g + b = 1$.

Considering that the human vision and display equipment have different sensitivities for different colors, the values of each ratio coefficient are different in the color matching. The green coefficient is quite big, the red coefficient is in middle, and the blue coefficient is minimum (Poynton, 1996). International standard Rec ITU-R BT.709 has standardized the corresponding CRT display coefficients for red, green, and blue. A linear combination formula of red, green, and blue is used to calculate the true brightness for the modern cameras and display devices:

$$Y_{709} = 0.2125\, R + 0.7154\, G + 0.0721\, B \tag{8.2}$$

Although the contribution of blue to brightness is the smallest, the human vision has a particularly good ability to distinguish blue color. If the number of bits assigned to the blue color is less than the number of bits assigned to the red color or green color, the blue region in the image may have a false contour effect.

In some color matching cases, adding only R, G, B together may not always produce the required color feeling. In this situation, add one of the three primary colors to the side of matched color (*i. e.*, written negative), in order to achieve equal color matching, such as

$$C \equiv rR + gG - bB \tag{8.3}$$
$$bB + C \equiv rR + gG \tag{8.4}$$

In the above discussion, the **color matching** refers to the same visual perception of the color. It is also known as color matching with "the same appearance heterogeneity" (metameric). In this case, there is no constrain for the spectral energy distribution of color.

8.1.3 Chroma and Chromaticity Diagram

In order to express the color, the concept of chroma is used. In order to represent the chroma, the representation method with chromaticity diagram is introduced.

8.1.3.1 Chroma and Chromatic Coefficient

People often use brightness, hue, and saturation to represent color properties. **Brightness** corresponds to the brilliance of the color. **Hue** is related to the wavelength of the main light in the spectrum, or it represents the main color that the viewer feels. **Saturation** is related to the purity of a certain hue, the pure spectral color is completely saturated, and the saturation is gradually reduced with the adding of the white light.

Hue and saturation together are called **chroma**. Color can be represented with both brightness and chroma. Let X, Y, and Z represent the three stimuli used to compose a certain color C, then the three stimulus values and CIE's R, G, and B have the following relationship:

$$\begin{bmatrix} X \\ Y \\ Z \end{bmatrix} = \begin{bmatrix} 0.4902 & 0.3099 & 0.1999 \\ 0.1770 & 0.8123 & 0.0107 \\ 0.0000 & 0.0101 & 0.9899 \end{bmatrix} \begin{bmatrix} R \\ G \\ B \end{bmatrix} \tag{8.5}$$

On the other side, according to the X, Y, Z stimulus values, three primary colors can also be obtained:

$$\begin{bmatrix} R \\ G \\ B \end{bmatrix} = \begin{bmatrix} 2.3635 & -0.8958 & -0.4677 \\ -0.5151 & 1.4264 & 0.0887 \\ 0.0052 & -0.0145 & 1.0093 \end{bmatrix} \begin{bmatrix} X \\ Y \\ Z \end{bmatrix} \tag{8.6}$$

For white light, it has $X = 1$, $Y = 1$, $Z = 1$. Let the amount of stimulation of each scale factor be x, y, z, there is $C = xX + yY + zZ$. Scale factor x, y, z are also known as chromatic coefficients,

$$x = \frac{X}{X + Y + Z} \quad y = \frac{Y}{X + Y + Z} \quad z = \frac{Z}{X + Y + Z} \tag{8.7}$$

From eq. (8.7), it can be seen that

$$x + y + z = 1 \tag{8.8}$$

8.1.3.2 Chromaticity Diagram

Simultaneously using three primary colors to represent a particular color needs to use 3-D space, so it will cause certain difficulty in mapping and display. To solve the problem, CIE developed a **chromaticity diagram** of tongue form (also called shark fin shape) in 1931, to project the three primary colors onto a 2-D chroma plane. By means of a chromaticity diagram, the proportion of the three primary colors to form a color may conveniently be presented by a 2-D composition.

A chromaticity diagram is shown in Figure 8.2 where the unit for wavelength is nm, the horizontal axis corresponds to red coefficient, the vertical axis corresponds to green coefficient, the blue coefficient value can be obtained by $z = 1 - (x + y)$, whose direction corresponds to come out from the inside of paper. The points on the tongue-shaped contour provide chromaticity coordinates corresponding to saturated colors. the spectrum corresponding to blue-violet is at the lower-left portion of chromaticity diagram, the spectrum corresponding to green is at the upper-left portion of chromaticity diagram, and the spectrum corresponding to red is at the lower-right portion

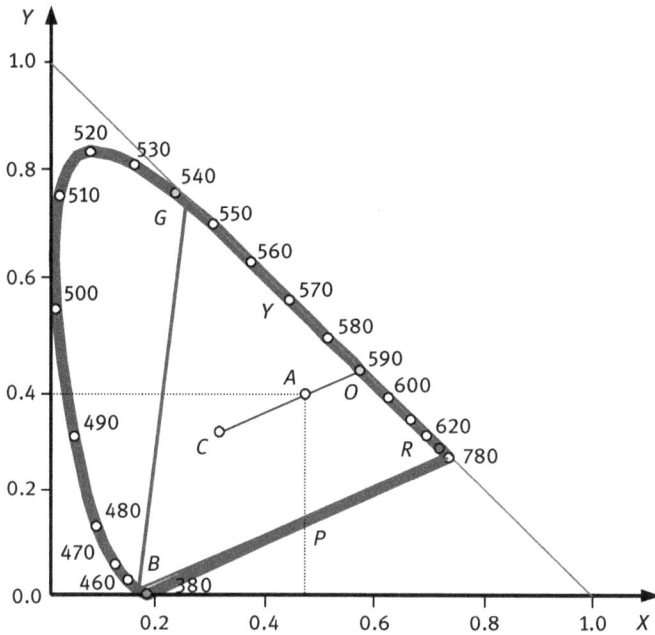

Figure 8.2: Illustration of a chromaticity diagram.

of chromaticity diagram. The tongue-shaped contour could be considered as the trajectory of a narrow spectrum, containing only a single wavelength of energy, passing through the range of 380 ~780 nm wavelength.

It should be noted that the straight line connecting 380 nm and 780 nm at the boundary of the chromaticity diagram corresponding to the purple series from blue to red, which is not available in the light spectrum. From the point of view of human vision, the feeling to purple does not generate only by a single wavelength, it requires the mixing of a shorter wavelength light and a longer wavelength light. In the chromaticity diagram, the line corresponding to the purple series connects the extreme blue (comprising only short-wavelength energy) and the extreme red (containing only long-wavelength energy).

8.1.3.3 Discussion on Chromaticity Diagram

By observing and analyzing chromaticity diagram, the following are noted:

(1) Each point in the chromaticity diagram corresponds to a visual perceived color. Conversely, any visible color occupies a determined position in the chromaticity diagram. For example, the chromaticity coordinates of the point A in Figure 8.2 are $x = 0.48$, $y = 0.40$. The point inside the triangle taking (0, 0), (0, 1), (1, 0) as the vertices, but outside the tongue-shaped contour, corresponds to invisible color.

(2)   The points on the tongue-shaped contour represent pure colors. When the points move toward the center, the mixed white light increases while the purity decreases. At the center point C, various spectral energies become equal. The combination of all three primary colors with one-third proportion will produce white, where the purity is zero. The color purity is generally referred to as the saturation of the color. In Figure 8.2, the point A is located in 66% of the distance from the point C to point of pure orange, so the saturation at point A is 66%.

(3)   In the chromaticity diagram, the two colors at the two endpoints of a straight line passing through the point C are complementary colors. For example, a non-spectral color in the purple section can be represented by the complementary color (C) at the other end of the line passing through the point C, and can be expressed as 510 C.

(4)   The points on the chromaticity diagram border have different hues. All points on the line connecting the center point C and a boundary point have the same hue. In Figure 8.2, a straight line is drawn from point C through point A to point O on the boundary (orange, about 590 nm), the dominant wavelength of point A is then 590 nm, the hue of point A is equal to that of point O.

(5)   In chromaticity diagram, all points on any line connecting two endpoints represent a new color that can be obtained by adding the colors represented by the two endpoints. To determine the color range that three colors can composite, connect the three points corresponding to the three colors into a triangle. For example, in Figure 8.2, any color whose corresponding point is located in the triangle with red, green, and blue points as vertex can be obtained by the three-color composition, and all colors outside the triangle cannot be obtained by the three-color composition. As the triangle formed by any given three points (corresponding to three fixed colors) cannot enclose all colors inside chromaticity diagram, it is not possible to use only three primary colors (single wavelength) to obtain all visible colors.

Example 8.1 Chromaticity triangles of PAL and NTSC systems

It is required for a variety of color display systems to select the appropriate R, G, B as the basic colors. For example, the chromaticity triangles of PAL and NTSC television systems in use are shown in Figure 8.3. Factors influencing the selection of three basic colors R, G, B for a system are:

(1)   From the technical point of view, it is difficult to produce highly saturated colors, so these basic colors are not fully saturated color;

(2)   It is better to make a bigger triangle with R, G, B as the vertices to include a larger area, that is, in containing more different colors;

(3)   The saturated cyan color is not commonly used. Therefore, in the chromaticity triangle, the red vertex is closest to fully saturation (spectrum border), while the green vertex and blue vertex have bigger distances from a totally saturated points (the NTSC system has more blue-green than that of the PAL system).

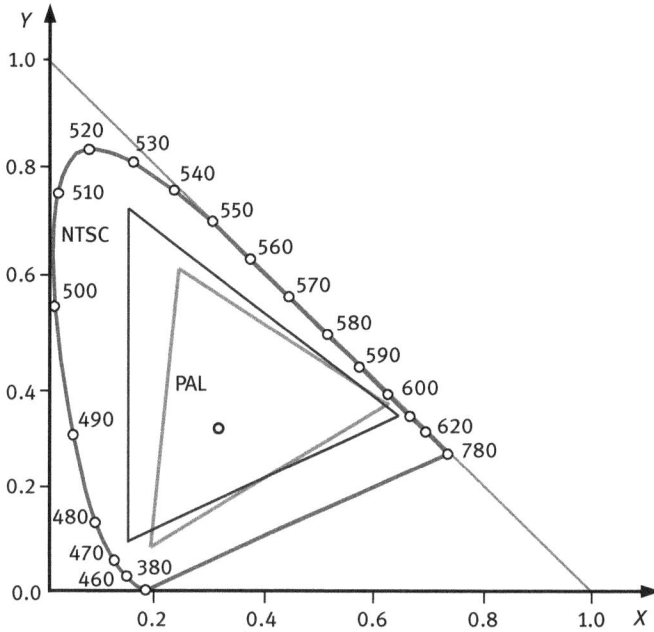

Figure 8.3: Chromaticity triangles of PAL and NTSC systems.

Back to the discussion of "metameric" in color match, it may be considered in the chromaticity diagram that the chromaticity coordinates of a color can only express its appearance but cannot express its spectral energy distribution.

8.2 Color Models

A **color model**, also called a **color space**, is a specification of a coordinate system in which each color is represented by a single point. To effectively express the color information, establishing and selecting suitable color representation models are needed. When creating a color model, since a color can be represented by three basic colors, it is required to build a 3-D spatial coordinate system, in which each point represents a particular kind of color. Various color models have been proposed.

From the application point of view, the proposed color models can be divided into two categories: one is for hardware devices, such as color printer or color display monitors; the other is for application-oriented visual perception or color processing and analysis, such as a variety of image techniques and the animation of color graphics. Some typical models from these two categories are introduced below.

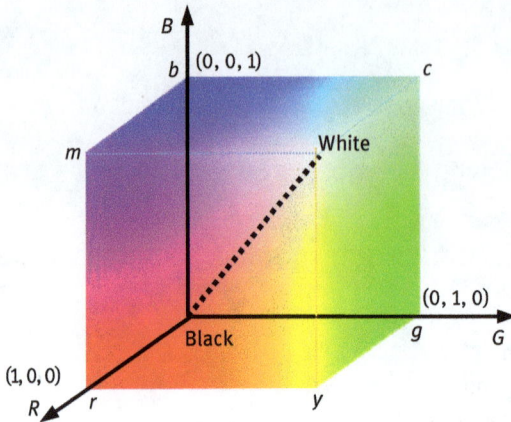

Figure 8.4: The RGB color cube.

8.2.1 Hardware-Orientated Models

The **hardware-orientated models** are suitable for image acquisition as well as image output and display applications.

8.2.1.1 RGB Model

This model is a popularly used color model, which is based on a Cartesian coordinate system, where the color space is represented by a cube as shown in Figure 8.4. The origin of coordinate system corresponds to black, the vertices being farthest from the origin correspond to white. In this model, the gray values from black to white are distributed along the line connecting these two vertices, while the remaining points within the cube corresponds to other different colors that can be represented by vectors. For convenience, the cube is generally normalized to unit cube so that all R, G, B values in **RGB color model** are in the interval of $[0, 1]$.

According to this model, each color image includes three independent color planes, or each color image can be decomposed into three plane. On the contrary, if one image can be represented by three planes, then using this model is quite convenient.

Example 8.2 Safe RGB color

True color RGB images are represented with 24-bits, that is, each R, G, B has 8 bits. The values of R, G, B are each quantized to 256 levels, the combination thereof may constitute more than 1,600 million colors. Actually, it is not needed to distinguish so many colors. Furthermore, this often makes too high requirements for display system. So a subset of color is devised and can be reliably displayed on various systems. This subset is called **safe RGB colors**, or the set of all-system-safe colors.

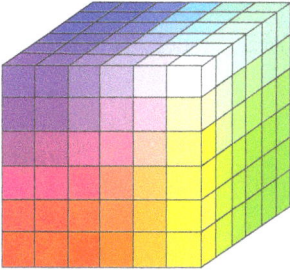

Figure 8.5: Safe RGB colors.

There are totally 256 colors in this subset. It is obtained by taking six values each from R, G, B. These six values are 0, 51, 102, 153, 204, and 255. If these values are expressed in hexagonal number system, they are 00, 33, 66, 99, CC, and FF. An illustration for this subset is shown in Figure 8.5.

8.2.1.2 CMY Model

In the RGB model, each color appears in its primary spectral components of red (R), green (G), and blue (B). The **CMY color model** is based on the combination of RGB to produce the primary colors of cyan (C), magenta (M), and yellow (Y):

$$C = 1 - R \tag{8.9}$$
$$M = 1 - G \tag{8.10}$$
$$Y = 1 - B \tag{8.11}$$

8.2.1.3 I_1, I_2, I_3 Model

This model is named according to its three components. The I_1, I_2, I_3 model is based on the experiments with natural image processing (for the purpose of segmentation). It is obtained by a linear transform of RGB, given by

$$I_1 = \frac{R + G + B}{3} \tag{8.12}$$
$$I_2 = \frac{R - B}{2} \tag{8.13}$$
$$I_3 = \frac{2G - R - B}{4} \tag{8.14}$$

A variation of the I_1, I_2, I_3 model is the I_1, I_2', I_3' model (Bimbo, 1999), in which

$$I_2' = R - B \tag{8.15}$$
$$I_3' = (2G - R - B)/2 \tag{8.16}$$

8.2.1.4 Normalized Model

A **normalized color model** is derived from the RGB model (Gevers and Smeulders, 1999), given by

$$l_1(R, G, B) = \frac{(R - G)^2}{(R - G)^2 + (R - B)^2 + (G - B)^2} \tag{8.17}$$

$$l_2(R, G, B) = \frac{(R - B)^2}{(R - G)^2 + (R - B)^2 + (G - B)^2} \tag{8.18}$$

$$l_3(R, G, B) = \frac{(G - B)^2}{(R - G)^2 + (R - B)^2 + (G - B)^2} \tag{8.19}$$

This model is invariant to the viewing direction, the object orientation, the lighting direction, and the brightness variation.

8.2.1.5 Color Model for Television

The **color model for TV** is also based on the combination of RGB. In the PAL system, the color model used is the **YUV model**, where Y denotes the brightness component and U and V are called chroma components and are proportional to color differences $B - Y$ and $R - Y$, respectively. YUV can be obtained from the normalized R', G', B' ($R' = G' = B' = 1$ corresponding to white) in the PAL system, given by

$$Y = 0.299R' + 0.587G' + 0.114B' \tag{8.20}$$

$$U = -0.147R' - 0.289G' + 0.436B' \tag{8.21}$$

$$V = 0.615R' - 0.515G' - 0.100B' \tag{8.22}$$

Reversely, R', G', B' can also be obtained from Y, U, V as

$$R' = 1.000Y + 0.000U + 1.140V' \tag{8.23}$$

$$G' = -1.000Y - 0.395U + 0.581V' \tag{8.24}$$

$$B' = 1.000Y + 2.032U + 0.001V \tag{8.25}$$

In the NTSC system, the color model used is the **YIQ model**, where Y denotes the brightness component, and I and Q are the results rotating the U and V by 33°. YIQ can be obtained from the normalized $R', G', B' (R' = G' = B' = 1$ corresponding to white) in the NTSC system, given by:

$$Y = 0.299R' + 0.587G' + 0.114B' Y = 0.299R' + 0.587G' + 0.114B' \tag{8.26}$$

$$I = 0.596R' - 0.275G' - 0.321B' I = 0.596R' - 0.275G' - 0.321B' \tag{8.27}$$

$$Q = 0.212R' - 0.523G' + 0.311B' Q = 0.212R' - 0.523G' + 0.311B' \tag{8.28}$$

Reversely, R', G', B' can also be obtained from Y, I, Q as

$$R' = 1.000Y + 0.956I + 0.620Q \tag{8.29}$$
$$G' = 1.000Y - 0.272I - 0.647Q \tag{8.30}$$
$$B' = 1.000Y - 1.108I + 1.700Q \tag{8.31}$$

8.2.1.6 Color Model for Video

One color model commonly used in video is the **YC_BC_R color model,** where Y represents the luminance component, and C_B and C_R represent chrominance components. The luminance component can be obtained by means of the RGB component of the color:

$$Y = rR + gG + bB \tag{8.32}$$

where r, g, b are proportional coefficients. The chrominance component C_B represents the difference between the blue portion and the luminance value, and the chrominance component C_R represents the difference between the red portion and the luminance value (so they are also called color difference components)

$$\begin{aligned} C_B &= B - Y \\ C_R &= R - Y \end{aligned} \tag{8.33}$$

In addition, there is $C_G = G - Y$, but it can be obtained from C_B and C_R. The inverse transformation from Y, C_B, C_R to R, G, B can be expressed as

$$\begin{bmatrix} R \\ G \\ B \end{bmatrix} = \begin{bmatrix} 1.0 & -0.00001 & 1.40200 \\ 1.0 & -0.34413 & -0.71414 \\ 1.0 & 1.77200 & 0.00004 \end{bmatrix} \begin{bmatrix} Y \\ C_B \\ C_R \end{bmatrix} \tag{8.34}$$

In the practical YC_BC_R color coordinate system, the value range of Y is [16, 235]; the value ranges of C_B and C_R are both [16, 240]. The maximum value of C_B corresponds to blue (C_B = 240 or $R = G = 0$, $B = 255$), and the minimum value of C_B corresponds to yellow (C_B = 16 or $R = G = 255$, $B = 0$). The maximum value of C_R corresponds to red (C_R = 240 or $R = 255$, $G = B = 0$), and the minimum value of C_R corresponds to cyan (C_R = 16 or $R = 0$, $G = B = 255$).

The spatial sampling rate of the video refers to the sampling rate of the luminance component Y, which is typically doubling the sampling rate of the chrominance components C_B and C_R. This could reduce the number of pixels per line, but does not change the number of lines per frame. This format is referred to as 4: 2: 2, which means that every four Y samples correspond to two C_B samples and two C_R samples.

The format with even lower data volume than the above format is 4: 1: 1 format, that is, each four Y sampling points corresponding to one C_B sample points and one C_R sample points. However, in this format the horizontal and vertical resolutions are very asymmetric. Another format with the same amount of data is the 4: 2: 0 format, which still corresponds to one C_B sample point and one C_R sample point for every four Y samples, but both C_B and C_R are sampled horizontally and vertically with the half of the sample rate. Finally, a 4: 4: 4 format is also defined for applications that require high resolution, that is, the sampling rate of the luminance component Y is the same as the sampling rates of the chrominance components C_B and C_R. The correspondence between the luminance and chrominance sampling points in the above four formats is shown in Figure 8.6.

8.2.2 Perception-Orientated Models

Perception-orientated color models are more suitable for describing colors in terms of human interpretation. Among them, the HSI (hue, saturation, and intensity) model is a basic model. Other models include the HCV (hue, chroma, and value) model, the HSV (hue, saturation, and value) model, the HSB (hue, saturation, and brightness) model, and the $L^*a^*b^*$ model (see below).

8.2.2.1 HSI Model

In the **HSI color model**, the intensity component is decoupled from the components carrying hue and saturation information. As a result, the HSI model is an ideal tool for developing image-processing algorithms based on the color descriptions that are natural and intuitive to humans.

The HSI space is represented by a vertical intensity axis and the locus of color points that lie on planes perpendicular to this axis. One of the planes is shown in Figure 8.7. In this plane, a color point P is represented by a vector from the origin to the color point P. The hue of this color is determined by an angle between a reference line (usually the line between the red point and the origin) and the color vector. The saturation of this color is proportional to the length of the vector.

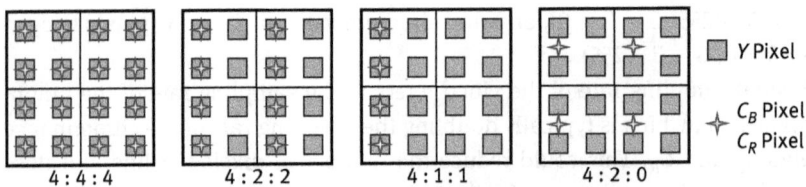

Figure 8.6: Illustration of four sample formats (two adjacent rows belonging to two different fields).

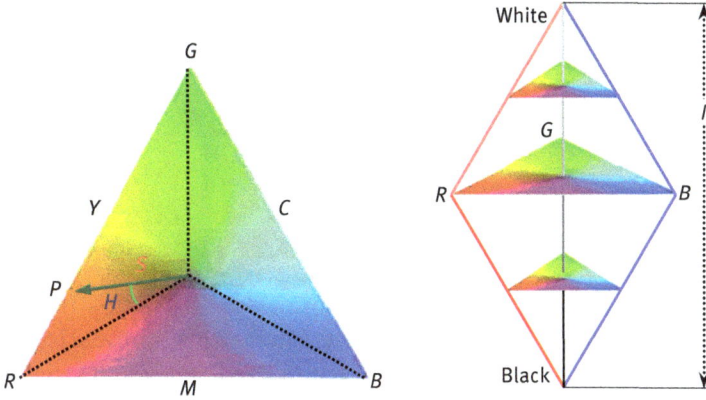

Figure 8.7: The HSI color plane and space.

Given an image in RGB color format, the corresponding HSI components can be obtained by

$$
H = \begin{cases} \arccos\left\{\dfrac{(R-G)+(R-B)}{2\sqrt{(R-G)^2+(R-B)(G-B)}}\right\} & R \neq G \text{ or } R \neq B \\[4mm] 2\pi - \arccos\left\{\dfrac{(R-G)+(R-B)}{2\sqrt{(R-G)^2+(R-B)(G-B)}}\right\} & B > G \end{cases} \tag{8.35}
$$

$$
S = 1 - \frac{3}{R+G+B}\min(R,G,B) \tag{8.36}
$$

$$
I = \frac{R+B+G}{3} \tag{8.37}
$$

On the other hand, given the H, S, I components, their corresponding R, G, B values are calculated in three ranges

(1) $H \in [0°, 120°]$

$$
B = I(1-S) \tag{8.38}
$$

$$
R = I\left[1 + \frac{S\cos H}{\cos(60° - H)}\right] \tag{8.39}
$$

$$
G = 3I - (B+R) \tag{8.40}
$$

(2) $H \in [120°, 240°]$

$$
R = I(1-S) \tag{8.41}
$$

$$
G = I\left[1 + \frac{S\cos(H - 120°)}{\cos(180° - H)}\right] \tag{8.42}
$$

$$
B = 3I - (R+G) \tag{8.43}
$$

(3) $H \in [240°, 360°]$

$$G = I(1 - S) \tag{8.44}$$

$$B = I\left[1 + \frac{S\cos(H - 240°)}{\cos(300° - H)}\right] \tag{8.45}$$

$$R = 3I - (G + B) \tag{8.46}$$

Example 8.3 Different color components of an image
Figure 8.8 shows different components of a color image. Figures 8.8(a), (b), and (c) are the R, G, B components of the image, respectively. Figures 8.8(d), (e), and (f) are the H, S, I components of the image, respectively.

◻

8.2.2.2 HSV Model
HSV color model is closer to human perception of color than the HSI model. The coordinate system of the HSV model is also a cylindrical coordinate system, but is generally represented by a hexcone (see Figure 8.9 of Plataniotis and Venetsano-poulos, 2000).

(a)   (b)   (c)

(d)   (e)   (f)

Figure 8.8: Different components of a color image.

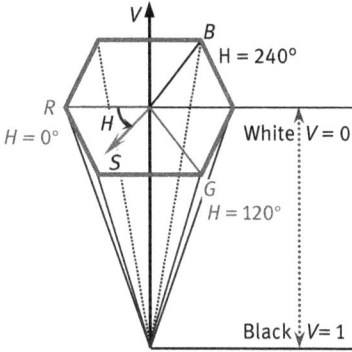

Figure 8.9: Coordinate system of HSV model.

The values of R, G, and B (all in $[0, 255]$) at a certain point in the RGB space can be converted to HSV space. The corresponding H, S, V values are:

$$H = \begin{cases} \arccos\left\{ \dfrac{(R-G) + (R-B)}{2\sqrt{(R-G)^2 + (R-B)(G-B)}} \right\} & B \le G \\[4mm] 2\pi - \arccos\left\{ \dfrac{(R-G) + (R-B)}{2\sqrt{(R-G)^2 + (R-B)(G-B)}} \right\} & B > G \end{cases} \tag{8.47}$$

$$S = \frac{\max(R, G, B) - \min(R, G, B)}{\max(R, G, B)} \tag{8.48}$$

$$V = \frac{\max(R, G, B)}{255} \tag{8.49}$$

8.2.2.3 HSB Model

The **HSB color model** is based on the opposite color theory (Hurvich and Jameson, 1957). Opposite color theory derives from the observation of opposing hues (red and green, yellow and blue), which counteract each other if the colors of opposing hues are superimposed. For a given frequency stimulus, the ratio of the four basic hues (red r, green g, yellow y, and blue b) can be deduced and a hue response equation can be established. It is also possible to deduce a non-hue response equation corresponding to the brightness sensed in a spectral stimulus. According to these two response equations, the hue coefficient function and the saturation coefficient function can be obtained. The hue coefficient function represents the ratio of the hue response of each frequency to all hue responses, and the saturation coefficient function represents the ratio of the hue response of each frequency to all hue-free responses. HSB model can explain many of the psychophysical phenomena about the color.

From the RGB model, and using the following linear transformation formula:

$$I = wb = R + G + B \tag{8.51}$$

$$rg = R - G \tag{8.52}$$

$$yb = 2B - R - G \tag{8.53}$$

In eq. (8.51), w and b are called white and black, respectively. In eq. (8.52) and eq. (8.53), rg and yb are the opposite color hues of the color space, respectively.

Although the opposite color model can be derived from the RGB model using a linear transformation formula, it is much more appropriate to model the perceived color than the RGB model.

8.2.2.4 $L^*a^*b^*$ Model

From the perspective of image processing, the description of color should be closer to the perception of color as possible. From the perspective of uniform perception, the distance between two perceived colors should be proportional to the distance between the two colors in the color space that express them. The $L^*a^*b^*$ **color model**, defined by CIE, is such a uniform color model in which the distance between two points in color space is proportional to the difference between the corresponding colors perceived by human eyes. The uniform color space model is essentially a color model that is visually perceptible, but more homogeneous in visual perception.

The $L^*a^*b^*$ color model is based on opposite color theory and reference white point (Wyszecki and Stiles, 1982). It is independent of devices used, and is suitable for applications with natural illumination. L^*, a^*, b^* are nonlinearly related to R, G, B, through the middle terms X, Y, Z (tri-stimulus):

$$X = 0.4902R + 0.3099G + 0.1999B \tag{8.54}$$
$$Y = 0.1770R + 0.8123G + 0.0107B \tag{8.55}$$
$$Z = 0.0000R + 0.0101G + 0.9899B \tag{8.56}$$

L^*, a^*, b^* are then computed by

$$L^* = \begin{cases} 116(Y/Y_0)^{1/3} - 16 & \text{if } Y/Y_0 > 0.008856 \\ 903.3(Y/Y_0)^{1/3} & \text{if } Y/Y_0 \le 0.008856 \end{cases} \tag{8.57}$$

$$a^* = 500\left[f(X/X_0) - f(Y/Y_0)\right] \tag{8.58}$$
$$b^* = 200\left[f(Y/Y_0) - f(Z/Z_0)\right] \tag{8.59}$$

where

$$f(t) = \begin{cases} t^{1/3} & \text{if } t > 0.008856 \\ 7.787t + 16/116 & \text{if } t \le 0.008856 \end{cases} \tag{8.60}$$

The $L^*a^*b^*$ color model covers all of the visible light spectrum and accurately expresses color in a variety of display, print and input devices. It has more emphasis on the green (more sensitive to green), followed by red and blue. But it does not provide a direct display format, so it must be converted to other color space to display.

8.3 Pseudo-Color Enhancement

Pseudo-color is also called **false color**. **Pseudo-color enhancements** consist of assigning colors to a gray-level image to enhance the distinction between different gray levels in the image.

In cases where the relative color in the image is more important than a specific expression of color, such as satellite images, microscopic image or X-ray images, using the pseudo-color technology to enhance image is a good choice (Umbaugh, 2005). Pseudo-color enhancement technology can be used either in the spatial domain or in the frequency domain.

In the pseudo-color image enhancement process, the input is a gray-level image, while the output is a colored image. Pseudo-color enhancement assigns the different gray-level regions in original image to diverse artificial colors in order to distinguish them more easily and clearly. This process is actually a color-painting (coloring) process. Three kinds of approaches for assigning pseudo-color according to the characteristics of gray-level regions in original images are presented in the following.

8.3.1 Intensity Slicing

The magnitude of an image can be considered a brightness function in 2-D coordinates. If a plane is placed parallel to the coordinate plane of the image, this plane will slice the brightness function (**intensity slicing**) and split this function into two gray-level sections. This process is shown in Figure 8.10.

According to Figure 8.8, for each input gray level, if it is under the slicing gray level l_m, it is assigned a specific color (here C_m); if it is above the slicing gray level l_m, it is assigned a different color (here C_{m+1}). In this way, the original gray-level image becomes an image with two different colors. The group of pixels whose original values are bigger than l_m and the group of pixels whose original values are smaller than l_m can be easily distinguished.

This method is also called pseudo-color mapping based on intensity slicing and can be extended to a more general situation. Suppose that M planes perpendicular to the intensity axis are defined at levels $l_1, l_2, \ldots, l_M$. Let level l_0 represent black ($f(x, y) = 0$), and level l_L represent white ($f(x, y) = L$). In the case of $0 < M < L$, the M planes divide the gray levels into $M + 1$ intervals. For each interval, a color can be assigned to it. This is given by

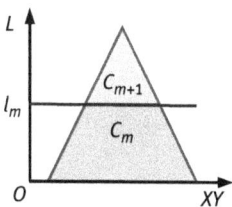

Figure 8.10: Illustration of intensity slicing.

$$f(x, y) = c_m \quad \text{if} \quad \begin{matrix} f(x, y) \in R_m \\ m = 0, 1, \cdots, M \end{matrix} \qquad (8.61)$$

where c_m is the color associated with the m-th intensity interval R_m defined by the partitioning planes at $l = m - 1$ and $l = m$.

Example 8.4 Pseudo-color mapping based on intensity slicing
The color assigned to a certain range of gray level is determined by three components; for each component different mapping functions can be designed. One example is given in Figure 8.11, in which the horizontal axis is for gray level. Figure 8.11(a) shows that the entire gray-level range [0, 1] was divided into four parts, each will be mapped by a different colors (0 to 1/4 is mapped by C_1, 1/4 ~ 2/4 is mapped by C_2, 2/4 to 3/4 is mapped by C_3, 3/4 ~ 4/4 is mapped by C_4). Let the three components of C_i are (R_i, G_i, B_i). The mapping functions in Figure 8.11(b) ~ Figure 8.11(d) are designed for R_i, G_i, B_i, respectively. Thus obtained color C_1 is dark blue, color C_2 is partial medium brown, color C_3 is partial light blue, and color C_4 is bright yellow, as shown in the circles at the left column of Figure 8.11(a). They will be converted into gray scale and displayed as black, dark gray, light gray, and white in grayscale print. ◻

8.3.2 Gray Level to Color Transformations

The idea underlying this approach of pseudo-color enhancement is to perform three independent transformations on the gray-level value of any input pixel. The three transformation functions are shown in Figure 8.12.

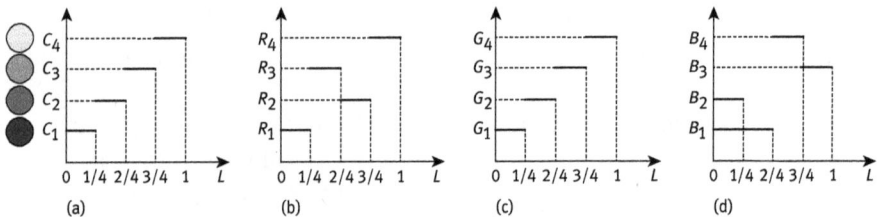

Figure 8.11: Pseudo-color mapping based on intensity slicing.

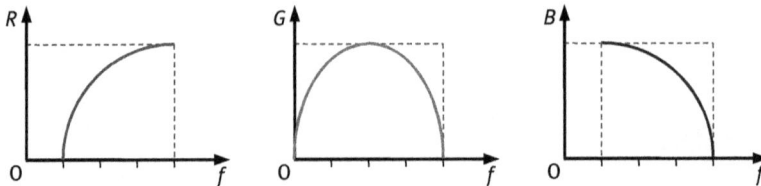

Figure 8.12: Pseudo-color transformation functions.

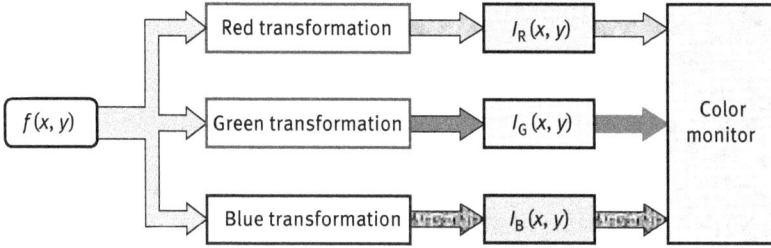

Figure 8.13: Process diagram of gray level to color transformations.

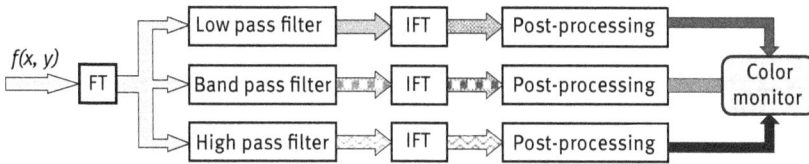

Figure 8.14: The flowchart frequency domain filtering for pseudo-color enhancement.

The three obtained transformation results are fed separately into the red, green, and blue channels of a color monitor. The whole process is shown in Figure 8.13.

8.3.3 Frequency Domain Filtering

Pseudo-color enhancement can also be performed with a variety of filters in the frequency domain. The basic idea of **frequency domain filtering** is to assign a particular color to the different regions of image, depending on the frequency content of each region. A basic flowchart is shown in Figure 8.14. The Fourier transform (FT) of the input image have been passed through three different filters (low-pass, band-pass, and high-pass filters, respectively, each of them covering the one-third of entire spectrum) to obtain different frequency components. For each of the frequency components within the range, the inverse Fourier transform (IFT) is performed to transform the component back to the image domain. The results can be further processed (*e. g.*, by histogram equalization, or by histogram specification). Putting the respective results into the red, green, and blue input channels of a color monitor will produce the final enhanced image.

8.4 Full-Color Enhancement

Pseudo-color enhancements change a gray-level image to a colored image, while the full-color enhancements directly treat color images, that is, with both input and output images are true color images.

8.4.1 Strategy and Principles

There are two strategies used to process full-color images. One is to process each component of a color image individually. This is a **per-color component approach**. In this case, the techniques for gray-level images can be used for each color component. The results for every component are then combined to give full-color image. Another strategy is to consider each pixel in a color image as having three values, or in other words, the magnitude of the image should be represented by a vector. This is a **vector-based approach**. In this case, a color pixel is represented by $C(x, y)$, then $C(x, y) = [R(x, y)G(x, y)B(x, y)]^T$.

Depending on the characteristics of the processing operation, the results of two processing strategies may be the same or may be different. To make the above two approaches equivalent, two conditions have to be satisfied. First, the process has to be applicable to both scalars and vectors. Second, the operation for each component of a vector must be independent of that for other components. One typical example of the process satisfying these two conditions is neighborhood averaging. For a gray-level image, neighborhood averaging summarizes all values of pixels covered by the averaging mask and divides this sum by the number of pixels. For a color image, neighborhood averaging can be done either by processing each property vector (vector computation) or by treating each component (scalar computation) first and then summarizing the results. This equivalence can be described by

$$\sum_{(x,y)\in N} C(x, y) = \sum_{(x,y)\in N} [R(x, y) + G(x, y) + B(x, y)]$$

$$= \left\{ \sum_{(x,y)\in N} R(x, y) + \sum_{(x,y)\in N} G(x, y) + \sum_{(x,y)\in N} B(x, y) \right\}$$

(8.62)

The above approach is also suitable for weighted averaging. Similarly, in the color image sharpening using Laplace operator the same results can be obtained

$$\nabla^2 [C(x, y)] = \begin{bmatrix} \nabla^2 [R(x, y)] \\ \nabla^2 [G(x, y)] \\ \nabla^2 [B(x, y)] \end{bmatrix}$$

(8.63)

So performing sharpening filter of color images could be either directly with vectors or first with each components and then combining these results. It should be noted that not all processing operations can be treated as above, in general only the linear processing operations will produce equivalent results in two cases.

8.4.2 Single Component Mapping

A color image can be seen as a combination of the three components of the color image. The enhancement of any one component of the color image would change the visual effects of the color image.

8.4.2.1 Basic Principle and Steps
The enhancement of color images, with transform methods, can be represented by

$$g_i(x, y) = T_i[f_i(x, y)] \quad i = 1, 2, 3 \tag{8.64}$$

where $[T_1, T_2, T_3]$ forms the set of mapping functions.

One image can be either decomposed into R, G, B components or decomposed into H, S, I components. In both cases, if linear transforms are used to increase the brightness of an image (the slope of the transformation line is k), it has

$$g_i(x, y) = kf_i(x, y) \quad i = 1, 2, 3 \tag{8.65}$$

Then in the RGB space, the process indicated in eq. (8.55) should be performed for all three components. Because the feeling of human eye for three components H, S, and I are relatively independent, so in the HSI space it is possible to improve the visual effect of a color image by only enhancing one of three color components of the image. In the HSI space, if only one component (here suppose the first component) needs to be considered, then a simple process would be given by

$$g_1(x, y) = kf_1(x, y) \tag{8.66}$$
$$g_2(x, y) = f_2(x, y) \tag{8.67}$$
$$g_3(x, y) = f_3(x, y) \tag{8.68}$$

In the above discussion, each transformation depends on only one component in the color space. If only the changes in one component are considered, then the grayscale image enhancement and frequency domain filtering enhancement (and restoration) methods introduced in Chapters 3 (and in Chapter 4) can be used directly. The main steps for enhancing a full-color image include:
(1) Transforming R, G, B components into H, S, I space.
(2) Enhancing one of the components using the techniques for gray-level images.
(3) Transforming the new H, S, I components back to R, G, B space.

In the following, the enhancements of each one of H, S, I components are discussed separately.

8.4.2.2 Intensity Enhancement

Enhancement of intensity can be carried out with the techniques for gray-level images directly (such as histogram transformation or gray-level mapping). The results will be the increase of visual details. **Intensity enhancement** will not change the color content of the original image, but the enhanced image can appear different in chroma. The reason is that the enhancement of the intensity, that is, the change of intensity, may make a different impression for hue or saturation to human eyes.

8.4.2.3 Saturation Enhancement

The enhancement of saturation is similar to the enhancement of intensity, but the operation is often simpler. For example, in **saturation enhancement**, multiplying the saturation component of an image by a factor greater than 1 could make the color more brilliant, while multiplying the saturation component of an image by a factor smaller than 1 could decrease the sensitivity of the color.

Example 8.5 Illustration of saturation enhancement

A set of images for the illustration of saturation enhancement results is given in Figure 8.15. Figure 8.15(a) is an original color image. Figure 8.15(b) is the result of the increasing saturation (multiplying the saturation component by a factor greater than 1), in which the color becomes more saturated, the contrast is increased, and the edges become clearer. Figure 8.15(c) is the result of reducing the saturation (multiplying the saturation component by a factor smaller than 1), in which the low-saturated regions in the original image almost lose their chroma components (becomes gray-level regions) and the whole image appears to be monotonous. ◎

8.4.2.4 Hue Enhancement

Hue enhancement has some particularity. According to the representation of the HSI model, hue corresponds to an angle and this angle has a module of 2π. The adjustment of the hue value (the angle value) will shift the color in the hue spectrum. If this modification is moderate, the color would become either more "warm" or more "cold." However, if this modification is considerable, the color will be radically changed.

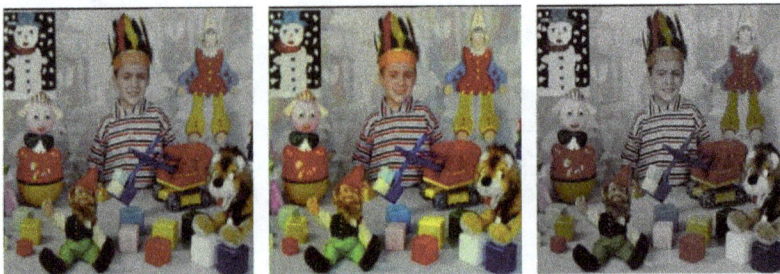

Figure 8.15: Illustration of saturation enhancement.

Figure 8.16: Illustration of saturation and hue enhancement.

Example 8.6 Illustration of hue enhancement

A set of images for the illustration of saturation and hue enhancement results is given in Figure 8.16. Figure 8.16(a) is the original color image. Figures 8.16(b) and (c) are the results of saturation enhancement, and the effects of the enhancements are similar to Figures 8.15(b) and (c), respectively. Figure 8.16(d) is the result of hue enhancement by subtracting a small value from the original hue value, in which red becomes violet while blue becomes green. Figure 8.16(e) is the result of hue enhancement by adding a large value to the original hue value, in which all the original colors are inversed, producing a result like an image negative (see Section 3.2.2). ▨

8.4.3 Full-Color Enhancements

Single component mapping has the advantage of simplicity as it separates the intensity, the saturation, and the hue. The disadvantage is that this mapping will change completely the perceptive feeling and such a change is difficult to control. Therefore, in some enhancement applications, all components should be taken into consideration.

8.4.3.1 Color Slicing

In color spaces (either RGB or HSI), each color occupies a fixed location in the space. For a natural image, the colors of the pixels corresponding to one object are often clustered. Consider an object region W in an image, if its corresponding cluster in the color space can be determined, this region will be distinguished from other regions and the enhancement purpose can be attained. This technique is parallel to the intensity slicing presented for pseudo-color enhancement and is called **color slicing**.

In the following, this technique is discussed using the RGB color space. For other color spaces, the procedure would be similar. Suppose the three color components corresponding to region W are $R_W(x, y)$, $G_W(x, y)$, and $B_W(x, y)$, respectively. First, their individual mean values (the coordinates of cluster center) are calculated by

$$m_R = \frac{1}{\#W} \sum_{(x,y)\in W} R_W(x,y) m_R = \frac{1}{\#W} \sum_{(x,y)\in W} R_W(x,y) \tag{8.69}$$

$$m_G = \frac{1}{\#W} \sum_{(x,y)\in W} G_W(x,y) \tag{8.70}$$

$$m_B = \frac{1}{\#W} \sum_{(x,y)\in W} B_W(x,y) \tag{8.71}$$

where $\#W$ denotes the number of pixels in region W. The next step is to determine the distribution width of d_R, d_G, and d_B. According to the mean values and the distribution width, the enclosing box in the color space for the corresponding pixels belonging to region W is given by $\{m_R - d_R/2 : m_R + d_R/2; m_G - d_G/2 : m_G + d_G/2; m_B - d_B/2 : m_B + d_B/2\}$. The enhancement can be performed in this box.

8.4.3.2 Color Linear Filtering

The above process is based on the point operation. The enhancement can also be done with the mask operation, which is called **color filtering**. Take neighborhood averaging as an example, which is a linear filtering process. Suppose the neighbors of a color pixel $C(x, y)$ are represented by W, then the result of the mask convolution is

$$C_{ave}(x,y) = \frac{1}{\#W} \sum_{(x,y)\in W} C(x,y) = \frac{1}{\#W} \begin{bmatrix} \sum_{(x,y)\in W} R(x,y) \\ \sum_{(x,y)\in W} G(x,y) \\ \sum_{(x,y)\in W} B(x,y) \end{bmatrix} \tag{8.72}$$

In other words, the color image is first decomposed into three gray-level images, the neighborhood averaging is then applied using the same masks for these images, and the obtained results are finally combined. This process is also suitable for the weighted averaging.

8.4.3.3 Color Median Filtering

Median filtering is based on data ranking or ordering. Since the color image is a vector image, and a vector is defined by both its magnitude and direction, the sorting of pixel values in the color image cannot be obtained directly by extending the ranking methods for the grayscale images. However, although there is no general and widely accepted method for general vector ranking without ambiguity, it is still possible to define a class of subordering methods for **color median filtering**, including the marginal ordering, conditional ordering, and reduced ordering.

In the **marginal ordering**, it is required to sort independently for each color component, and the final ranking is based on the pixel values of the same ordinal value of all components. In the **conditional ordering**, selecting one of the components and taking it as a scalar for ordering (if the values from the component are same, then a second component can also be considered, or even the third component could be considered), the pixel value of color images would be sorted in accordance with the order of this component. In the **reduced ordering**, all the pixel values (vector values) are first converted into scalar quantity by using some given simplification functions. However, attention should be paid to the interpretation of the results, as the ordering of vector value cannot be explained in the same way as the ordering of scalar value. For example, if a similarity function or distance function is used as a simplification function and the results are ranked in ascending order, then in the ranking results the data in the front of data set would be close to "center," while the outfield point is at the end of the data set.

Example 8.7 An example of subordering methods
Given a set of $N = 7$ colored pixels: $f_1 = [7, 1, 1]^T, f_2 = [4, 2, 1]^T, f_3 = [3, 4, 2]^T, f_4 = [6, 2, 6]^T, f_5 = [5, 3, 5]^T, f_6 = [3, 6, 6]^T, f_7 = [7, 3, 7]^T$. The ranking result obtained by marginal ordering is ($\Rightarrow$ indicates the ranking result):

$$\{7, 4, 3, 6, 5, 3, 7\} \Rightarrow \{3, 3, 4, 5, 6, 7, 7\}$$
$$\{1, 2, 4, 2, 3, 6, 3\} \Rightarrow \{1, 2, 2, 3, 3, 4, 6\}$$
$$\{1, 1, 2, 6, 5, 6, 7\} \Rightarrow \{1, 1, 2, 5, 6, 6, 7\}$$

Then, the ranked vectors are: $f_1 = [3, 1, 1]^T, f_2 = [3, 2, 1]^T, f_3 = [4, 2, 2]^T, f_4 = [5, 3, 5]^T, f_5 = [6, 3, 6]^T, f_6 = [7, 4, 6]^T, f_7 = [7, 6, 7]^T$. The median vector is $[5, 3, 5]^T$.

The ranking result obtained by conditional ordering is (using the third component for ranking): $f_1 = [7, 1, 1]^T, f_2 = [4, 2, 1]^T, f_3 = [3, 4, 2]^T, f_4 = [5, 3, 5]^T, f_5 = [6, 2, 6]^T, f_6 = [3, 6, 6]^T, f_7 = [7, 3, 7]^T$. The median vector is also $[5, 3, 5]^T$.

Now, considering the reduced ordering if the distance function is taken as the simplification function, that is, if $r_i = \{[f - f_m]^T[f - f_m]\}^{1/2}$, where $f_m = \{f_1 + f_2 + f_3 + f_4 + f_5 + f_6 + f_7\}/7 = [5, 3, 4]^T$, then the result obtained by distance computation is: $r_1 = 4.12, r_2 = 3.61, r_3 = 3, r_4 = 2.45, r_5 = 1, r_6 = 4.12, r_7 = 3.61$. The ranked vectors are: $f_1 = [5, 3, 5]^T, f_2 = [6, 2, 6]^T, f_3 = [3, 4, 2]^T, f_4 = [4, 2, 1]^T, f_5 = [7, 3, 7]^T, f_6 = [7, 1, 1]^T, f_7 = [3, 6, 6]^T$. The median vector is also $[5, 3, 5]^T$.

Under the above conditions of a group of color pixels, the median vectors obtained by using three subordering methods are the same, that is $[5, 3, 5]^T$. In addition, this median vector is also one of the original vector. However, it is not always the case in real applications. For example, if the first vector from $f_1 = [7, 1, 1]^T$ to $f_1 = [7, 1, 7]^T$ is changed, that is, only one of the component is modified, then the result would be quite different. The median vector obtained by the marginal ordering would be $[5, 3, 6]^T$, which is not an original vector. The median vector obtained by the conditional ordering would be $[6, 2, 6]^T$, which is different from the above case though is nevertheless an original vector. The median vector obtained by the reduced ordering would be $[5, 3, 5]^T$, which is still same as the above case. In these circumstances, the three ordering methods give the totally different results. ◎

Besides, it should be noted that the above three ordering methods have different characteristics. For example, the result from the marginal ordering does not often have the one-to-one correspondence with the original data, such as the obtained median vector might be not any of the original vectors. On the other side, the three components of a vector are not treated equally in conditional ordering, as only one of the three components is considered in ranking, so there will be problems caused by bias.

Finally, based on the reduced ordering, a method called **"vector ranking"** can be obtained with the help of similarity measure. The following quantity is computed:

$$R_i(\boldsymbol{f}_i) = \sum_{j=1}^{N} s(\boldsymbol{f}_i, \boldsymbol{f}_j) \tag{8.73}$$

where N is the number of vectors, $s(\cdot)$ is a similarity function (can be defined by distance). The ranking is conducted according to the value of $R_i = R(\boldsymbol{f}_i)$. Here the consideration is focused on the internal relationship among vectors. The output of a vector median filter obtained according to "vector ranking" is a vector that has the minimum sum distance with a set of vectors, it is the first vector if this set of vectors are ranked in ascending order. The norm used by vector median filter is dependent on the type of noise encountered. If the noise is correlated with the image, then L_2 norm should be used, while if the noise is not correlated with the image, then L_1 norm should be used.

Example 8.8 Color median filtering
A set of illustrations obtained with colored median filtering, by using the three types of vector ordering methods described above, to eliminate noise are shown in Figure 8.17. Figure 8.17(a) is an original image. Figure 8.17(b) is the result of adding 10% correlated salt-and-pepper noise. Figure 8.17(c) is the result obtained by using a median filter based on marginal ordering. Figure 8.17(d) is the difference image of Figures 8.17(a) and (c). Figure 8.17(e) is the result obtained by using a median filter based on conditional ordering. Figure 8.17(f) is the difference image of Figures 8.17(a) and (e). Figure 8.17(g) is the result obtained by using a median filter based on reduced ordering. Figure 8.17(h) is the difference image of Figures 8.17(a) and (g).

It is seen from Figure 8.15 that the result based on reduced ordering is better than the result based on conditional ordering, and the result based on conditional ordering is better than the result based on marginal ordering. However, each type of ranking method has several concrete techniques, whose effects on noise removing depend also on the images used. This example only takes one of these techniques, and cannot be considered as covering all cases. ◻

8.4.3.4 False Color Enhancement
False color enhancement can be considered as a special type of full-color enhancement. The input and output of false color enhancement are both color images (same as in full-color enhancement), so it differs from pseudo-color enhancement. In the false

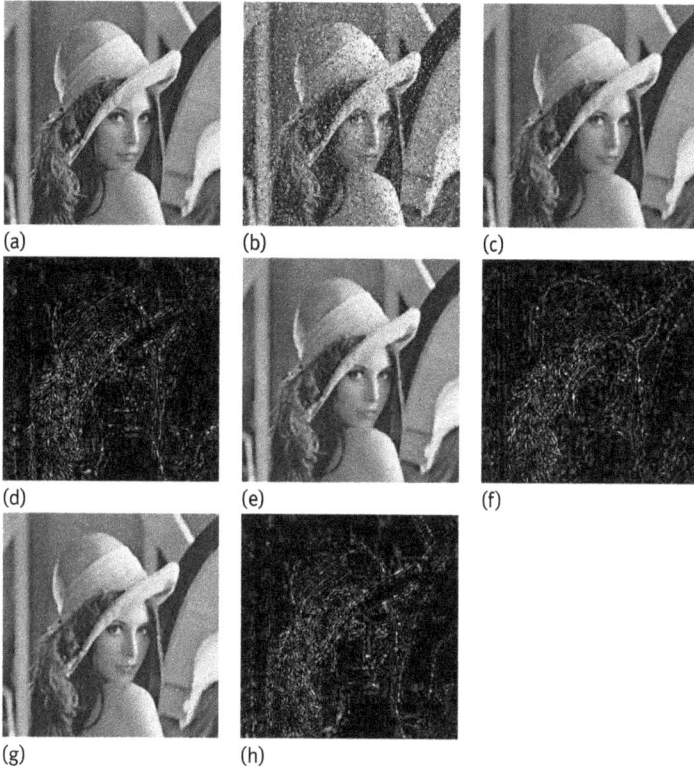

Figure 8.17: Illustration of color median filtering.

color enhancement, the values of each pixel in original color image will be mapped linearly or nonlinearly to different location in the same or different color spaces in considering the enhancement purpose. According to the necessity, some interest parts of the original color image could be made with completely different colors that could be also false colors quite unlike the people's expectations (even unnatural), which can make these image parts easier to get attention. For example, if a particular tree from a row of trees is required to introduce, then its leaves can be converted from the green color to the red color in order to attract more attention of observers.

If using R_O, G_O, and B_O to represent the R, G, and B components of an original image, and using R_E, G_E, and B_E to represent the R, G, and B components of an enhanced image, then a linear method of false color enhancement can be written as:

$$\begin{bmatrix} R_E \\ G_E \\ B_E \end{bmatrix} = \begin{bmatrix} m_{11} & m_{12} & m_{13} \\ m_{21} & m_{22} & m_{23} \\ m_{31} & m_{32} & m_{33} \end{bmatrix} \begin{bmatrix} R_O \\ G_O \\ B_O \end{bmatrix} \tag{8.74}$$

For example, the following simple operation could map green to red, blue to green and red to blue:

$$\begin{bmatrix} R_E \\ G_E \\ B_E \end{bmatrix} = \begin{bmatrix} 0 & 1 & 0 \\ 0 & 0 & 1 \\ 1 & 0 & 0 \end{bmatrix} \begin{bmatrix} R_O \\ G_O \\ B_O \end{bmatrix}$$ (8.75)

False color enhancement could make use of the property of human eye that has different sensitivity for various lights of distinct wavelength. For example, mapping red-brown wall brick to green would improve discernment details. In practice, if sensor used in the image acquisition phase is sensitive to the visible light spectrum and nearby spectrum (such as infrared or ultraviolet), false color enhancement is essential for the distinction of difference light sources in display.

8.5 Problems and Questions

8-1* There are many ways to achieve the same visual effects as the direct mixing of the three primary colors of light, for example:

(1) Rapidly projecting the three primary colors of light, in a certain order, onto the same surface.

(2) Projecting the three primary colors of light separately onto three close points in the same surface.

(3) Let the two eyes observe the different colors of one same image.

Which properties of the human visual system have been used in each of these methods?

8-2 Compare the locations of the points marked A, B, and D in the chromaticity diagram in Figure Problem 8-2, and discuss their color characteristics.

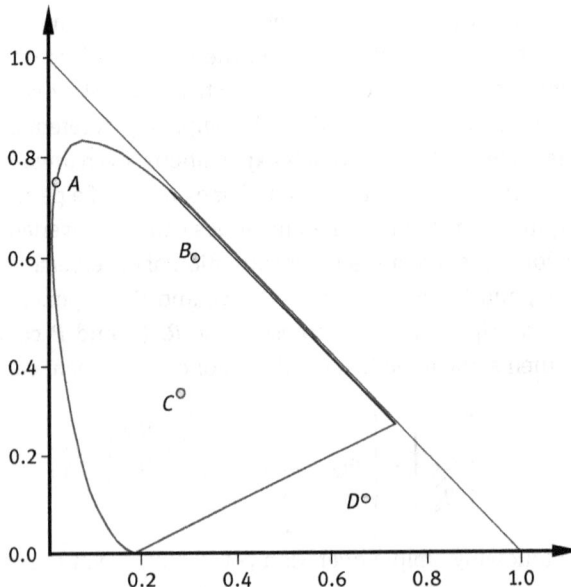

Figure Problem 8-2

8-3 In the RGB color cube:

(1) Mark the locations of all points with a gray value of 0.5.

(2) Mark the positions of all points with a saturation value of 1.

8-4 In the HSI double pyramid structure:

(1) Mark the locations of all points with a brightness value of 0.25 and a saturation value of 0.5.

(2) Mark the locations of all points with a brightness value of 0.25 and a hue value of 0.5 (the angle is expressed in radians).

8-5 This book discusses the HS plane in HSI space using a triangular area. Some other books have used a circular area, where the HSI space is cylindrical or spindered. Compare the characteristics of the two representations. Which other shape areas can you still select? Why?

8-6 Proving the formula for calculating S in eq. (8.33) holds for all points in the HSI color triangle in Figure 8.6(a).

8-7 The points in the RGB space have a corresponding relationship with the points in the HSV space:

(1) Calculate the H, S, I values of the six vertices r, g, b, c, m, y of the RGB color cube in Figure 8.4.

(2) Mark the letter or text representing the above six vertices at the corresponding positions of the HSI color entity shown in Figure 8.6(b).

8-8* Given two points A and B with the coordinates $(0.2, 0.4, 0, 6)$ and $(0.3, 0.2, 0.1)$ in the RGB coordinate system, respectively. Discuss the relationship between the HSI coordinates of point $C = A + B$ and the HSI coordinates of A or B.

8-9 Consider a photograph taken with a black-and-white camera in the early years for the national map, the gray levels of various provinces are different. Now, the three pseudo-color enhancement methods described in Section 8.3 can be used to convert the grayscale photograph into pseudo-color photograph. What will be the particularities or characteristics of each of the three results?

8-10 Discuss the results obtained with the saturation enhancement method and the hue enhancement method. What is alteration of the contrast in the result image compared to the contrast in the original image?

8-11 How to make the operation of image negative for a color image? Draw the mapping curves in the RGB and HSI spaces, respectively.

8-12 Analyze the principles of marginal ordering, conditional ordering, and reduced ordering. Discuss:

(1) When are the results of marginal ordering and conditional ordering different? Give a concrete numerical example.

(2) When are the results of marginal ordering and reduced ordering different? Give a concrete numerical example.

(3) When are the results of conditional ordering and reduced ordering different? Give a concrete numerical example.

8.6 Further Reading

1. **Color Vision and Chromaticity Diagram**
   - Further information on color vision and imaging can be found in Wyszecki and Stiles (1982), Palus (1998), MarDonald and Luo (1999), Plataniotis and Venetsanopoulos (2000), and Koschan and Abidi (2010).
   - Color can not only stimulate the senses but also play an important role for people to understand the content and atmosphere of the scene (Xu and Zhang, 2005).

2. **Color Models**
   - In the HSI model, a series of 2-D color pictures can be obtained by projecting the color space onto one of the axes, see Liu and Zhang (1998).

3. **Pseudo-Color Enhancement**
   - More discussion on false color can be found in Pratt (2007).

4. **Full-Color Enhancement**
   - More and more attentions have been paid on research of full-color image processing technology in recent years. For further information, see Pratt (2007), Gonzalez and Woods (2008), Koschan and Abidi (2010).
   - A comparative study of vector sorting in color image processing can be found in Liu and Zhang (2010).

Answers to Selected Problems and Questions

Chapter 1 Introduction to Image Processing

1-4 (1) $256 \times 256 \times (1 + 8 + 1)/9{,}600 = 68.3$ s.

(2) $1024 \times 1024 \times [3(1 + 8 + 1)]/38{,}400 = 819.2$ s.

1-8 The minimum gray level is 0 (every position is 0), the maximum gray level is 12 (every position is 3), during which the gray levels can be continuous, so this template can represent a total of 13 gray levels.

Chapter 2 Image Acquisition

2-3 The camera coordinates are $(-0.2, -0.4, 0)$. The image plane coordinates are $(-0.2, -0.4)$.

2-9 The output luminous flux of the street lamp is

$$I = \frac{\Phi}{\Omega} = \frac{\Phi}{4\pi}$$

The luminance at the distance r is

$$B = \frac{I}{S} = \frac{I}{4\pi r^2} = \frac{\Phi}{(4\pi r)^2}$$

Taking the values of r into the above formula, we get

$$B_{50} = \frac{\Phi}{(4\pi r)^2} = 5.066 \times 10^{-3} \text{cd/m}^2$$

$$B_{100} = \frac{\Phi}{(4\pi r)^2} = 1.267 \times 10^{-3} \text{cd/m}^2$$

Chapter 3 Image Enhancement

3-2 One method is first to move the image up, down, left, and right one pixel, respectively, to obtain four separated translation images. Then the four resulted images are undergone an "exclusive-OR" operation, to obtain the borders of the object in four directions. Finally, the contours are obtained by ANDing all these border segments.

3-6 The computation results are listed in Table Solution 3-6.

Table: Solution 3-6

Label	Computation	Steps and results							
1	Original gray level k	0	1	2	3	4	5	6	7
2	Original histogram s_k	0.174	0.088	0.086	0.08	0.068	0.058	0.062	0.384
3	Original cumulative histogram	0.174	0.262	0.348	0.428	0.496	0.554	0.616	1.00
4	Specified histogram		0.4			0.2			0.4
5	Specified cumulative histogram		0.4			0.6			1.0
6S	SML mapping	1	1	1	1	1	4	4	7
7S	Determine mapping relation		0, 1, 2, 3, 4 → 1				5, 6, → 4		7 → 7
8S	Resulted histogram		0.496			0.12			0.384
6G	GML mapping	1	1	1	1	4	4	4	7
7G	Determine the mapping relation		0, 1, 2, 3 → 1				4, 5, 6 → 4		7 → 7
8G	Resulted histogram		0.428			0.188			0.384

Note: Steps 6S to 8S are for SML mapping. Steps 6G to 8G are for GML mapping.

Chapter 4 Image Restoration

4-2 If the degradation system H in Figure 4.5 is linear and position-invariant, eq. (4.8) can be written as $g(x, y) = h(x, y) \otimes f(x, y) + n(x, y)$. Taking its Fourier transform for both sides gives $G(u, v) = F(u, v)H(u, v) + N(u, v)$. By seeking the square of the modulus (note that f and n are not relevant), $|G(u, v)|^2 = |H(u, v)|^2|F(u, v)|^2 + |N(u, v)|^2$ is proved.

4-8 According to eq. (4.90), the transfer function is

$$R(u, v) = \frac{H^*(u, v)}{|H(u, v)|^2 + s\,|P(u, v)|^2}$$

where $H(u, v)$ can be obtained by Fourier transform of $h(r) = h(x, y)$. Taking a square image, that is, $M = N$, get

$$H(u, v) = \frac{1}{M}\sum_{y=0}^{M-1}\sum_{x=0}^{M-1}\frac{x^2 + y^2 - 2\sigma^2}{\sigma^4}\exp\left[-\frac{x^2+y^2}{2\sigma^2}\right]\exp[-j2\pi(ux+vy)/M]$$

Further,

$$H^*(u, v) = \frac{1}{M}\sum_{y=0}^{M-1}\sum_{x=0}^{M-1}\frac{x^2 + y^2 - 2\sigma^2}{\sigma^4}\exp\left[-\frac{x^2+y^2}{2\sigma^2}\right]\exp[j2\pi(ux+vy)/M]$$

$P(u, v)$ can be obtained by using eq. (4.79):

$$P(u, v) = \frac{1}{M} \sum_{y=0}^{M-1} \sum_{x=0}^{M-1} p(x, y) \exp[-j2\pi(ux + vy)/M]$$

$$= \frac{1}{M}\{-\exp[-j2\pi v/M] - \exp[-j2\pi u/M] - \exp[-j2\pi(u + 2v)/M]$$

$$- \exp[-j2\pi(2u + v)/M] + 4\exp[-j2\pi(u + v)/M]\}$$

Substituting $H(u, v)$, $H^*(u, v)$, and $P(u, v)$ into the transfer function, the required constrained least square filter is obtained.

Chapter 5 Image Reconstruction from Projections

5-7 The proof can be expressed as: Prove that the Fourier transform of the image $f(x, y)$ projected on a straight line that has an angle θ to the X-axis is a section of the Fourier transform of $f(x, y)$ at the heading angle θ. Without loss of generality, it is assumed that $\theta = 0$.

Suppose the projection of image $f(x, y)$ on the X-axis is $g_y(x)$, then it has

$$g_y(x) = \int_{-\infty}^{+\infty} f(x, y)dy$$

Its Fourier transform is

$$G_y(u) = \int_{-\infty}^{+\infty} g_y(x) \exp[-j2\pi ux]dy = \int_{-\infty}^{+\infty}\int_{-\infty}^{+\infty} f(x, y) \exp[-j2\pi ux]dxdy$$

On the other side, the Fourier transform of $f(x, y)$ is

$$F(u, v) = \int_{-\infty}^{+\infty}\int_{-\infty}^{+\infty} f(x, y) \exp[-j2\pi(ux + vy)]dxdy$$

Let the value of v in the above formula to be 0, then $G_y(u) = F(u, 0)$, so the projection theorem for Fourier transform is proved.

5-10 According to eq. (5.31), $d(x, y)$ can be represented by

$$d(x, y) = \int_{0}^{\pi}\int_{-1}^{1} g(s, \theta)h(x \cos\theta + y \sin\theta - s)dsd\theta$$

$$= \int_{0}^{\pi}\int_{-1}^{1} \ln\left(\frac{I_r}{I(s, \theta)}\right) h(x \cos\theta + y \sin\theta - s)dsd\theta$$

where $I(s, \theta)$ represents the intensity of the ray passing through the object from the origin at a distance s and with an angle θ from the normal to the X-axis. Thus, the required CT value can be calculated as:

$$CT = k \left[\frac{\int_0^\pi \int_{-1}^1 \ln\left(\frac{I_r}{I(s,\theta)}\right) h(x\cos\theta + y\sin\theta - s)\,ds\,d\theta}{d_w(x,y)} - 1 \right]$$

Chapter 6 Image Coding

6-1 $\mathrm{SNR_{ms}} = 35$, $\mathrm{SNR_{rms}} = 5.92$, $\mathrm{SNR} = 8.5$ dB, PSNR $= 19.8$ dB.

6-12 Decision levels: $s_0 = 0$, $s_1 = k$, $s_2 = \infty$. Reconstruction levels: $t_1 = k/3$, $t_2 = 5k/3$.

Chapter 7 Image Watermarking

7-2 One example is the embedding validity, which means the probability of detecting the watermark immediately after the embedding of the watermark. Some watermarks are not easy to be embedded into the image due to fidelity requirements, which can result in less than 100% embedded efficiency. Another example is the false alarm rate, which refers to the probability of detecting out the watermark from the actually no watermark medium.

7-4 The first three.

Chapter 8 Color Image Processing

8-1 (1) Visual inertia (the basis of time-color mixing method).

  (2) The resolution limitation of human eye (the basis of spatial-color mixing method).

  (3) Physiological blending.

8-8 The point C is at $(0.5, 0.6, 0.7)$ of the RGB coordinate system, so from $I_A = (0.2 + 0.4 + 0.6)/3 = 0.4$, $I_B = (0.3 + 0.2 + 0.1)/3 = 0.2$, and $I_{A+B} = (0.5 + 0.6 + 0.7)/3 = 0.6$, it can be seen that $I_{A+B} = I_A + I_B$. Similarly, from $S_A = 0.5$, $S_B = 0.5$, and $S_{A+B} = 0.167$, it can be seen that the saturations S_A, S_B, and S_{A+B} do not have the superimposed relationship. In the same way, from $H_A = 193.33°$, $H_B = 33.33°$, and $H_{A+B} = 166.67°$, it can be seen that the hues H_A, H_B, and H_{A+B} also do not have the superimposed relationship.

References

[Alleyrand 1992] Alleyrand M R. 1992. Handbook of Image Storage and Retrieval Systems. Multiscience Press, New York.

[Andreu et al. 2001] Andreu J P, Borotsching H, Ganster H, et al. 2001. Information fusion in image understanding. Digital Image Analysis – Selected Techniques and Applications, Kropatsch W G, Bischof H (eds.). Springer, Heidelberg.

[Ashe 2010] Ashe M C. 2010. Image Acquisition and Processing with LabVIEW (2nd Ed.). CRC Press, Eleserv.

[ASM 2000] ASM International. 2000. Practical Guide to Image Analysis. ASM International, Materials Park, OH.

[Aumont 1997] Aumont J. 1997. The Image. British Film Institute, London. {Translation: Pajackowska C}.

[Ballard and Brown 1982] Ballard D H, Brown C M. 1982. Computer Vision. Prentice-Hall, New Jersey.

[Barber 2000] Barber D C. 2000. Electrical impedance tomography. The Biomedical Engineering Hand Book (2nd Ed.), CRC Press, Boca Raton (Chapter 68).

[Barni et al. 2001] Barni M, Bartolini F, Piva A. 2001. Improved wavelet-based watermarking through pixel-wise masking. IEEE-IP, 10(5): 783–791.

[Barni et al. 2003a] Barni M, et al. 2003a. What is the future for watermarking? (Part 1). IEEE Signal Processing Magazine, 20(5): 55–59.

[Barni et al. 2003b] Barni M, et al. 2003b. What is the future for watermarking? (Part 2). IEEE Signal Processing Magazine, 20(6): 53–58.

[Barnsley 1988] Barnsley M F. 1988. Fractals Everywhere. Academic Press, New York.

[Beddow 1997] Beddow J K. 1997. Image Analysis Sourcebook. American Universities Science and Technology Press, Iowa City.

[Bertalmio et al. 2001] Bertalmio M, Bertozzi A L, Sapiro G. 2001. Navier–Stokes, fluid dynamics, and image and video inpainting. Proc. CVPR, 417–424.

[Bertero and Boccacci 1998] Bertero M, Boccacci P. 1998. Introduction to Inverse Problems in Imaging. IOP Publishing Ltd, Philadelphia.

[Bimbo 1999] Bimbo A. 1999. Visual Information Retrieval. Morgan Kaufmann, Inc., Burlington.

[Bovik 2005] Bovik A. (Ed.). 2005. Handbook of Image and Video Processing (2nd Ed.) Elsevier, Amsterdam.

[Bow 2002] Bow S T. 2002. Pattern Recognition and Image Preprocessing (2nd Ed.). Marcel Dekker, Inc., New York.

[Boyd 1983] Boyd R W. 1983. Radiometry and the Detection of Optical Radiation. Wiley, New Jersey.

[Bracewell 1995] Bracewell R N. 1995. Two-Dimensional Imaging. Prentice Hall, New Jersey.

[Branden and Farrell 1996] Branden C J, Farrell J E. 1996. Perceptual quality metric for digitally coded color images. Proc. EUSIPCO-96, 1175–1178.

[Castleman 1996] Castleman K R. 1996. Digital Image Processing. Prentice-Hall, New Jersey.

[Censor 1983] Censor Y. 1983. Finite series-expansion reconstruction methods. Proceedings of IEEE, 71: 409–419.

[Chan and Shen 2001] Chan T F, Shen J H. 2001. Non-texture inpainting by curvature-driven diffusions (CDD). Journal of Visual Communication and Image Representation, 12(4): 436–449.

[Chan and Shen 2005] Chan T F, Shen J. 2005. Image Processing and Analysis – Variational, PDE, Wavelet, and Stochastic Methods. USA Philadelphia: Siam.

[Chen and Chen 2001] Chen J L, Chen X H. 2001. Special Matrices. Tsinghua University Press, Beijing.

[Chui 1992] Chui C K. 1992. An Introduction to WAVELETS. Academic Press, New York.

[Committee 1994]   Committee on the Encyclopedia of Mathematics. 1994. Encyclopedia of Mathematics. Science Press, Dordrecht.

[Committee 1996]   Committee on the Mathematics and Physics of Emerging Dynamic Biomedical Imaging. 1996. Mathematics and Physics of Emerging Biomedical Imaging. National Academic Press, Washington, D.C.

[Cosman et al. 1994]   Cosman P C, et al. 1994. Evaluating quality of compressed medical images: SNR, subjective rating, and diagnostic accuracy. Proceedings of IEEE, 82: 919–932.

[Cox et al. 2002]   Cox I J, Miller M L, Bloom J A. 2002. Digital Watermarking. Elsevier Science, Amsterdam.

[Criminisi et al. 2003]   Criminisi A, Perez P, Toyama K. 2003. Object removal by exemplar-based image inpainting. Proc. ICCV, 721–728.

[Davies 2012]   Davies E R. 2012. Computer and Machine Vision: Theory, Algorithms, Practicalities (4th Ed.). Elsevier, Amsterdam.

[Dean et al. 1995]   Dean T, Allen J, Aloimonos Y. 1995. Artificial Intelligence: Theory and Practice. Addison Wesley, New Jersey.

[Deans 2000]   Deans S R. 2000. Radon and Abel transforms. The Transforms and Applications Handbook (2nd Ed.) CRC Press, Boca Raton (Chapter 8).

[Dougherty and Astola 1994]   Dougherty E R, Astola J. 1994. An Introduction to Nonlinear Image Processing. SPIE Optical Engineering Press, Bellingham, Washington USA.

[Duan and Zhang 2010]   Duan F, Zhang Y-J. 2010. A highly effective impulse noise detection algorithm for switching median filters. IEEE Signal Processing Letters, 17(7): 647–650.

[Duan and Zhang 2011]   Duan F, Zhang Y-J. 2011. A parallel impulse-noise detection algorithm based on ensemble learning for switching median filters. Proc. Parallel Processing for Imaging Applications (SPIE-7872), 78720C-1-78720C-12.

[Edelman 1999]   Edelman S. 1999. Representation and Recognition in Vision. MIT Press, Boston.

[Faugeras 1993]   Faugeras O. 1993. Three-Dimensional Computer Vision: A Geometric Viewpoint. MIT Press, Boston.

[Finkel and Sajda 1994]   Finkel L H, Sajda P. 1994. Constructing visual perception. American Scientist, 82(3): 224–237.

[Forsyth 2003]   Forsyth D, Ponce J. 2003. Computer Vision: A Modern Approach. Prentice Hall, New Jersey.

[Forsyth 2012]   Forsyth D, Ponce J. 2012. Computer Vision: A Modern Approach (2nd Ed.). Prentice Hall, New Jersey.

[Fridrich and Miroslav 1999]   Fridrich J, Miroslav G. 1999. Comparing robustness of watermarking techniques. SPIE, 3657: 214–225.

[Fu 1987]   Fu K S, Gonzalez R C, Lee C S G. 1987. Robotics: Control, Sensing, Vision, and Intelligence. McGraw-Hill, New York.

[Furht et al. 1995]   Furht B, Smoliar S W, Zhang H J. 1995. Video and image processing in multimedia systems. Kluwer Academic Publishers, Dordrecht, 226–270.

[Gersho 1982]   Gersho A. 1982. On the structure of vector quantizer. IEEE-IT, 28: 157–166.

[Gevers and Smeulders 1999]   Gevers T, Smeulders A. 1999. Color-based object recognition. PR, 32(3): 453–464.

[Gibson 1950]   Gibson J J. 1950. The Perception of the Visual World. Houghton Mifflin, Boston.

[Girod 1989]   Girod B. 1989. The information theoretical significance of spatial and temporal masking in video signals. SPIE, 1077: 178–187.

[Gonzalez and Wintz 1987]   Gonzalez R C, Wintz P. 1987. Digital Image Processing (2nd Ed.). Addison-Wesley, New Jersey.

[Gonzalez and Woods 1992]   Gonzalez R C, Woods R E. 1992. Digital Image Processing (3rd Ed.). Addison-Wesley, New Jersey.

[Gonzalez and Woods 2002]   Gonzalez R C, Woods R E. 2002. Digital Image Processing, (2nd Ed.). Prentice Hall, New Jersey.

[Gonzalez and Woods 2008]   Gonzalez R C, Woods R E. 2008. Digital Image Processing (3rd Ed.). Prentice Hall, New Jersey.

[Gordon 1974]   Gordon R. 1974. A tutorial on ART (Algebraic Reconstruction Techniques). IEEE-NS, 21: 78–93.

[Goswami and Chan 1999]   Goswami J C, Chan A K. 1999. Fundamentals of Wavelets – Theory, Algorithms, and Applications. John Wiley & Sons, Inc., New Jersey.

[Hanson 1978]   Hanson A R, Riseman E M. 1978. Computer Vision Systems. Academic Press, Washington, D.C.

[Haralick 1992]   Haralick R M, Shapiro L G. 1992. Computer and Robot Vision, Vol. 1. Addison-Wesley, New Jersey.

[Haralick 1993]   Haralick R M, Shapiro L G. 1993. Computer and Robot Vision, Vol. 2. Addison-Wesley, New Jersey.

[Hartley 2000]   Hartley R, Zisserman A. 2000. Multiple View Geometry in Computer Vision. Cambridge University Press, Cambridge.

[Hartley and Zisserman 2004]   Hartley R, Zisserman A. 2004. Multiple View Geometry in Computer Vision (2nd Ed.). Cambridge University Press, Cambridge.

[Herman 1980]   Herman G T. 1980. Image Reconstruction from Projection – The Fundamentals of Computerized Tomography. Academic Press, Inc., New Jersey.

[Herman 1983]   Herman G T. 1983. The special issue on computerized tomography. Proceedings of IEEE, 71: 291–292.

[Horn 1986]   Horn B K P. 1986. Robot Vision. MIT Press, Boston.

[Huang 1965]   Huang T S. 1965. PCM Picture transmission. IEEE Spectrum, 2: 57–63.

[Huang 1993]   Huang T, Stucki P (Eds.). 1993. Special section on 3-D modeling in image analysis and synthesis. IEEE-PAMI, 15(6): 529–616.

[Hurvich and Jameson 1957]   Hurvich L M, Jameson D. 1957. An opponent process theory of colour vision. Psychological Review, 64(6): 384–404.

[Jain 1989]   Jain A K. 1989. Fundamentals of Digital Image Processing. Prentice-Hall, New Jersey.

[Jain 1995]   Jain R, Kasturi R, Schunck B G. 1995. Machine Vision. McGraw-Hill Companies. Inc., New York.

[Jähne 1997]   Jähne B. 1997. Digital Image Processing – Concepts, Algorithms and Scientific Applications. Springer, Heidelberg.

[Jähne 2004]   Jähne B. 2004. Practical Handbook on Image Processing for Scientific and Technical Applications (2nd Ed.). CRC Press, Boca Raton.

[Jähne and Haußecker 2000]   Jähne B, Haußecker H. 2000. Computer Vision and Applications: A Guide for Students and Practitioners. Academic Press, New Jersey.

[Jähne et al. 1999a]   Jähne B, Haußecker H, Geißler P. 1999a. Handbook of Computer Vision and Applications – Volume 1: Sensors and Imaging. Academic Press, New Jersey.

[Jähne et al. 1999b]   Jähne B, Haußecker H, Geißler P. 1999b. Handbook of Computer Vision and Applications – Volume 2: Signal Processing and Pattern Recognition. Academic Press, New Jersey.

[Jähne et al. 1999c]   Jähne B, Haußecker H, Geißler P. 1999c. Handbook of Computer Vision and Applications – Volume 3: Systems and Applications. Academic Press, New Jersey.

[Joyce 1985]   Joyce Loebl. 1985. Image Analysis: Principles and Practice. Joyce Loebl, Ltd, Gateshead.

[Julesz 1960]   Julesz B. 1960. Binocular depth perception of computer generated patterns. Bell System Technical Journal, 39, 1125–1162.

[Kak 1995]   Kak A. 1995. Editorial. CVIU, 61(2): 153–153.

[Kak and Slaney 2001]   Kak A C, Slaney M. 2001. Principles of Computerized Tomographic Imaging. Society for Industrial and Applied Mathematics, Philadelphia.

[Karnaukhov et al. 2002]   Karnaukhov V N, Aizenberg I N, Butakoff C, et al. 2002. Neural network identification and restoration of blurred images. SPIE, 4875: 303–310.

[Kitchen and Rosenfeld 1981]   Kitchen L, Rosenfeld A. 1981. Edge evaluation using local edge coherence. IEEE-SMC, 11(9): 597–605.

[Koschan and Abidi 2010] Koschan A, Abidi M. 2010. Digital Color Image Processing. Tsinghua University Press, Beijing. {Translation: Zhang Y-J}.

[Kropatsch 2001] Kropatsch W G, Bischof H (Eds.). 2001. Digital Image Analysis – Selected Techniques and Applications. Springer, Heidelberg.

[Kutter and Petitcolas 1999] Kutter M, Petitcolas F A P. 1999. A fair benchmarking for image watermarking systems. SPIE, 3657: 226–239.

[Kutter and Hartung 2000] Kutter M, Hartung F. 2000. Introduction to watermarking techniques. In: Information Hiding Techniques for Steganography and Digital Watermarking, Katzenbeisser S, Petitcolas F A (eds.). Artech House, Inc., Norwood. (Chapter 5).

[Lau and Arce 2001] Lau D L, Arce G R. 2001. Digital halftoning. In: Nonlinear Image Processing, Mitra S (ed.). Academic Press, New Jersey. (Chapter 13).

[Levine 1985] Levine M D. 1985. Vision in Man and Machine. McGraw-Hill, New York.

[Lewitt 1983] Lewitt R M. 1983. Reconstruction algorithms: transform methods. Proceedings of IEEE, 71: 390–408.

[Li and Zhang 2003] Li R, Zhang Y-J. 2003 A hybrid filter for the cancellation of mixed Gaussian noise and impulse noise, Proc. 4PCM, 1: 508–512.

[Libbey 1994] Libbey R L. 1994. Signal and Image Processing Sourcebook. Van Nostrand Reinhold, New York.

[Liu and Zhang 1998] Liu Z W, Zhang Y-J. 1998. Color image retrieval using local accumulative histogram. Journal of Image and Graphics, 3(7): 533–537.

[Liu and Zhang 2010] Liu K, Zhang Y-J. 2010. Color image processing based on vector ordering and vector filtering. Proc. 15NCIG, 51–56.

[Lohmann 1998] Lohmann G. 1998. Volumetric Image Analysis. John Wiley & Sons and Teubner Publishers, New Jersey.

[Luo et al. 2006] Luo S W, et al. 2006. Information Processing Theory of Visual Perception. Publishing House of Electronics Industry.

[Luryi 2014] Luryi S, Mastrapasqua M. 2014. Charge injection devices. In: Encyclopedia of Electrical and Electronics Engineering, John Wiley & Sons Inc., New Jersey. 258–263.

[Mahdavieh 1992] Mahdavieh Y, Gonzalez R C. 1992. Advances in Image Analysis. DPIE Optical Engineering Press.

[Marchand and Sharaiha 2000] Marchand-Maillet S, Sharaiha Y M. 2000. Binary Digital Image Processing – A Discrete Approach. Academic Press, New Jersey.

[MarDonald and Luo 1999] MarDonald L W, Luo M R. 1999. Colour Imaging – Vision and Technology. John Wiley & Sons Ltd, New Jersey.

[Marr 1982] Marr D. 1982. Vision – A Computational Investigation into the Human Representation and Processing of Visual Information. W.H. Freeman, New York.

[Mitra and Sicuranza 2001] Mitra S K, Sicuranza G L. (Eds). 2001. Nonlinear Image Processing. Academic Press, New Jersey.

[Moretti et al. 2000] Moretti B, Fadili J M, Ruan S, et al. 2000. Phantom-based performance evaluation: Application to brain segmentation from magnetic resonance images. Medical Image Analysis, 4(4): 303–316.

[Olsen 1993] Olsen S I. 1993. Estimation of noise in images: An evaluation. CVGIP-GMIP, 55(4): 319–323.

[Palus 1998] Palus H. 1998. Representations of colour images in different colour spaces. In: The Colour Image Processing Handbook, Sangwine S J, Horne R E N (eds.). Chapman & Hall, London.

[Pavlidis 1982] Pavlidis T. 1982. Algorithms for Graphics and Image Processing. Computer Science Press, Berlin-Heidelberg-New York.

[Pavlidis 1988] Pavlidis T. 1988. Image analysis. Annual Review of Computer Science, 3: 121–146.

[Plataniotis and Venetsanopoulos 2000] Plataniotis K N, Venetsanopoulos A N. 2000. Color Image Processing and Applications. Springer, Heidelberg.

[Poynton 1996]   Poynton C A. 1996. A Technical Introduction to Digital Video. John Wiley & Sons Inc., New Jersey.

[Pratt 2001]   Pratt W K. 2001. Digital Image Processing. John Wiley & Sons Inc., New Jersey.

[Pratt 2007]   Pratt W K. 2007. Digital Image Processing: PIKS Scientific Inside (4th Ed.). Wiley Interscience, New Jersey.

[Prince 2012]   Prince S J D. 2012. Computer Vision – Models, Learning, and Inference. Cambridge University Press, Cambridge.

[Rabbani and Jones 1991]   Rabbani M, Jones P W. 1991. Digital Image Compression Techniques. SPIE Press, Bellingham.

[Ritter and Wilson 2001]   Ritter G X, Wilson J N. 2001. Handbook of Computer Vision Algorithms in Image Algebra. CRC Press, Boca Raton.

[Rosenfeld and Kak 1976]   Rosenfeld A, Kak A C. 1976. Digital Picture Processing. Academic Press, New Jersey.

[Rosenfeld 1984]   Rosenfeld A. 1984. Image analysis: Problems, progress and prospects. PR, 17: 3–12.

[Russ 2002]   Russ J C. 2002. The Image Processing Handbook (4th Ed.). CRC Press, Boca Raton.

[Russ 2016]   Russ J. C. 2016. The Image Processing Handbook (7th Ed.). CRC Press, Boca Raton.

[Said and Pearlman 1996]   Said A, Pearlman W A. 1996. A new fast and efficient image codec based on set partitioning in hierarchical trees. IEEE-CSVT, 6(6): 243–250.

[Salomon 2000]   Salomon D. 2000. Data Compression: The Complete Reference (2nd Ed.). Springer – Verlag, Heidelberg.

[Serra 1982]   Serra J. 1982. Image Analysis and Mathematical Morphology. Academic Press, New Jersey.

[Shapiro and Stockman 2001]   Shapiro L, Stockman G. 2001. Computer Vision. Prentice Hall, New Jersey.

[Shen et al. 2009]   Shen B, Hu W, Zhang Y M, et al. 2009. Image inpainting via sparse representation. Proc. ICASSP, 697–700.

[Sheppand Logan 1974]   Shepp L A. and Logan B F. 1974. The Fourier reconstruction of a head section. IEEE-NS, 21: 21–43.

[Shih 2013]   Shih F Y (Ed.). 2013. Multimedia Security: Watermarking, Steganography, and Forensics. CRC Press, Boca Raton.

[Shirai 1987]   Shirai Y. 1987. Three-Dimensional Computer Vision. Springer-Verlag, Heidelberg.

[Snyder and Qi 2004]   Snyder W E, Qi H. 2004. Machine Vision. Cambridge University Press, Cambridge.

[Sonka et al. 1999]   Sonka M, Hlavac V, Boyle R. 1999. Image Processing, Analysis, and Machine Vision (2nd Ed.). Brooks/Cole Publishing.

[Sonka et al. 2008]   Sonka M, Hlavac V, Boyle R. 2008. Image Processing, Analysis, and Machine Vision (3rd Ed.). Thomson Learning, Totonto.

[Sweldens 1996]   Sweldens W. 1996. The lifting scheme: A custom-design construction of biorthogonal wavelets. Journal of Applied and Computational Harmonic Analysis, 3(2): 186–200.

[Szeliski 2010]   Szeliski R. 2010. Computer Vision: Algorithms and Applications. Springer, Heidelberg.

[Tekalp 1995]   Tekalp A M. 1995. Digital Video Processing. Prentice-Hall, New Jersey.

[Tsai 1987]   Tsai R Y. 1987. A versatile camera calibration technique for high-accuracy 3D machine vision metrology using off-the shelf TV camera and lenses. Journal of Robotics and Automation, 3(4): 323–344.

[Tsai and Zhang 2010]   Tsai W, Zhang Y-J. 2010. A frequency sensitivity-based quality prediction model for JPEG images. Proc. 5ICIG, 28–32.

[Umbaugh 2005]   Umbaugh S E. 2005. Computer Imaging – Digital Image Analysis and Processing. CRC Press, Boca Raton.

[Unzign 1997]   UnZign watermark removal software. 1997. http://altern.org/watermark/

[Wang and Zhang 2011]   Wang Y X, Zhang Y-J. 2011. Image inpainting via weighted sparse non-negative matrix factorization. Proc. International Conference on Image Processing, IEEE, 3470–3473.

[Wang et al. 1991]   Wang G, Lin T H, Cheng P C, et al. 1991. Scanning cone-beam reconstruction algorithms for X-ray micro-tomography. SPIE, 1556: 99.

[Wang et al. 2002]   Wang Y, Ostermann J, Zhang Y Q. 2002. Video Processing and Communications. Prentice Hall, New Jersey.

[Wang et al. 2005]   Wang Z M, Zhang Y-J, Wu J H. 2005. A wavelet domain watermarking technique based on human visual system. Journal of Nanchang University (Natural Science), 29(4): 400–403.

[Wei et al. 2005]   Wei Y C, Wang G, Hsieh J. 2005. Relation between the filtered back-projection algorithm and the back-projection algorithm in CT. IEEE Signal Processing Letters, 12(9): 633–636.

[Wyszecki and Stiles 1982]   Wyszecki G, Stiles W S. 1982. Color Science, Concepts and Methods, Quantitative Data and Formulas. John Wiley, New Jersey.

[Xu and Zhang 2005]   Xu F, Zhang Y-J. 2005. Atmosphere-based image classification through illumination and hue. SPIE, 5960: 596–603.

[Young and Renswoude 1988]   Young I T, Renswoude J. 1988. Three-dimensional image analysis. SPIN Program, 1–25.

[Young 1993]   Young I T. 1993. Three-dimensional image analysis. Proc. VIP'93, 35–38.

[Young et al. 1995]   Young I T, Gerbrands J J, Vliet L J. 1995. Fundamental of Image Processing. Delft University of Technology, The Netherlands.

[Zakia 1997]   Zakia R D. 1997. Perception and Imaging. Focal Press, Waltham, Massachusetts.

[Zeng 2009]   Zeng G S L. 2009. Medical Image Reconstruction: A Conceptual Tutorial. Higher Education Press, Beijing.

[Zhang 1992]   Zhang Y-J. 1992. Improving the accuracy of direct histogram specification. IEE EL, 28(3): 213–214.

[Zhang 1996a]   Zhang Y-J. 1996a. Image engineering in China: 1995. Journal of Image and Graphics, 1(1): 78–83.

[Zhang 1996b]   Zhang Y-J. 1996b. Image engineering in China: 1995 (Supplement). Journal of Image and Graphics, 1(2): 170–174.

[Zhang 1996c]   Zhang Y-J. 1996c. Image engineering and bibliography in China. Technical Digest of International Symposium on Information Science and Technology, 158–160.

[Zhang 1997a]   Zhang Y-J. 1997a. Image engineering in China: 1996. Journal of Image and Graphics, 2(5): 336–344.

[Zhang et al. 1997b]   Zhang Y-J, Yao Y R, He Y. 1997b. Automatic face segmentation using color cues for coding typical videophone scenes. SPIE, 3024: 468–479.

[Zhang 1998]   Zhang Y-J. 1998. Image engineering in China: 1997. Journal of Image and Graphics, 3(5): 404–414.

[Zhang 1999a]   Zhang Y-J. 1999a. Image engineering in China: 1998. Journal of Image and Graphics, 4(5): 427–438.

[Zhang et al. 1999b]   Zhang Y-J, Li Q, Ge J H. 1999b. A computer assisted instruction courseware for "Image Processing and Analysis". Proc. ICCE'99, 371–374.

[Zhang 1999c]   Zhang Y-J. 1999c. Image Engineering (1): Image Processing and Analysis. Tsinghua University Press.

[Zhang 2000a]   Zhang Y-J. 2000a. Image engineering in China: 1999. Journal of Image and Graphics, 5(5): 359–373.

[Zhang 2000b]   Zhang Y-J. 2000b. Image Engineering (2): Image Understanding and Computer Vision. Tsinghua University Press.

[Zhang 2001a]   Zhang Y-J. 2001a. Image engineering in China: 2000. Journal of Image and Graphics, 6(5): 409–424.

[Zhang *et al*. 2001b]   Zhang Y-J, Chen T, Li J. 2001b. Embedding watermarks into both DC and AC components of DCT. SPIE, 4314: 424–435.

[Zhang *et al*. 2001c]   Zhang Y-J, Liu W J, Zhu X Q, *et al*. 2001c. The overall design and prototype realization of web course "Image Processing and Analysis". In: Signal and Information Processing Techniques, Zeng Y F (ed.), Electronic Industry Press, Beijing, 452–455.

[Zhang 2002a]   Zhang Y-J. 2002a. Image engineering in China: 2001. Journal of Image and Graphics 7(5): 417–433.

[Zhang 2002b]   Zhang Y-J. 2002b. Image Engineering (3): Teaching References and Problem Solutions. Tsinghua University Press, Beijing.

[Zhang 2002c]   Zhang Y-J. 2002c. Image engineering and related publications. International Journal of Image and Graphics, 2(3): 441–452.

[Zhang *et al*. 2002d]   Zhang Y-J, Jiang F, Hu H J, *et al*. 2002d. Design and implementation of the self-test module in web courses. Proc. ICCE, 1092–1094.

[Zhang and Liu 2002e]   Zhang Y-J, Liu W J. 2002e. A new web course – "Fundamentals of Image Processing and Analysis". Proceedings of the 6th Global Chinese Conference on Computer in Education, 1: 597–602.

[Zhang 2003]   Zhang Y-J. 2003. Image engineering in China: 2002. Journal of Image and Graphics, 8(5): 481–498.

[Zhang 2004a]   Zhang Y-J. 2004a. Image Engineering in China: 2003. Journal of Image and Graphics, 9(5): 513–531.

[Zhang 2004b]   Zhang Y-J. 2004b. A combined structure suitable for web courses. New Horizon in Web-Based Learning, Cheung R, Lau R, Li Q (eds.), 189–195.

[Zhang 2004c]   Zhang Y-J. 2004c. Mapping laws for histogram processing of digital images. Journal of Image and Graphics, 9(10): 1265–1268.

[Zhang 2004d]   Zhang Y-J. 2004d. On the design and application of an online web course for distance learning. International Journal of Distance Education Technologies, 2(1): 31–41.

[Zhang 2004e]   Zhang Y-J. 2004e. Web Course of Image Processing and Analysis. High Education Press, Beijing.

[Zhang 2005a]   Zhang Y-J. 2005a. Image engineering in China: 2004. Journal of Image and Graphics, 10(5): 537–560.

[Zhang 2005b]   Zhang Y-J. 2005b. Better use of digital images in teaching and learning. Encyclopedia of Distance Learning, Idea Group Reference, Hershey PA, 1: 152–158.

[Zhang 2005c]   Zhang Y-J. 2005c. Image Engineering (2): Image Analysis (2nd Ed.). Tsinghua University Press, Beijing.

[Zhang 2005d]   Zhang Y-J. 2005d. Important design considerations for online web courses in distance learning. Encyclopedia of Distance Learning, Idea Group Reference, Hershey PA, 3: 1042–1047.

[Zhang 2006a]   Zhang Y-J. 2006a. Image engineering in China: 2005. Journal of Image and Graphics, 11(5): 601–623.

[Zhang 2006b]   Zhang Y-J. 2006b. Image Engineering (1): Image Processing (2nd Ed.). Tsinghua University Press, Beijing.

[Zhang 2007a]   Zhang Y-J. 2007a. Image engineering in China: 2006. Journal of Image and Graphics, 12(5): 753–775.

[Zhang 2007b]   Zhang Y-J. 2007b. A teaching case for a distance learning course: Teaching digital image processing. Journal of Cases on Information Technology, 9(4): 30–39.

[Zhang 2007c]   Zhang Y-J. 2007c. Image Engineering (2nd Ed.). Tsinghua University Press, Beijing.

[Zhang 2007d]   Zhang Y-J. 2007d. Image Engineering (3): Image Understanding (2nd Ed.). Tsinghua University Press.

[Zhang 2008a]   Zhang Y-J. 2008a. Image engineering in China: 2007. Journal of Image and Graphics, 13(5): 825–852.

[Zhang 2008b] Zhang Y-J. 2008b. A study of image engineering. In: Encyclopedia of Information Science and Technology (2nd Ed.), Information Science Reference, Hershey PA, Vol. VII, 3608–3615.

[Zhang 2008c] Zhang Y-J. 2008c. On the design and application of an online web course for distance learning. In: Handbook of Distance Learning for Real-Time and Asynchronous Information Technology Education, Information Science Reference, Hershey PA, 228–238 (Chapter 12).

[Zhang and Zhao 2008d] Zhang Y-J, Zhao X M. 2008d. Development of net courseware for "Image Processing". Proc. 14NCIG, 790–794.

[Zhang 2009a] Zhang Y-J. 2009a. Image engineering in China: 2008. Journal of Image and Graphics, 14(5): 809–837.

[Zhang 2009b] Zhang Y-J. 2009b. Teaching and learning image courses with visual forms. Encyclopedia of Distance Learning (2nd Ed.), IGI Global, Hershey PA, 4: 2044–2049.

[Zhang 2010] Zhang Y-J. 2010. Image engineering in China: 2009. Journal of Image and Graphics, 15(5): 689–722.

[Zhang 2011a] Zhang Y-J. 2011a. Image engineering in China: 2010. Journal of Image and Graphics, 16(5): 693–702.

[Zhang 2011b] Zhang Y-J. 2011b. A net courseware for "Image Processing". Proc. 6ICCGI, 143–147.

[Zhang 2012a] Zhang Y-J. 2012a. Image engineering in China: 2011. Journal of Image and Graphics, 17(5): 603–612.

[Zhang 2012b] Zhang Y-J. 2012b. Image Engineering (1): Image Processing (3rd Ed.). Tsinghua University Press, Beijing.

[Zhang 2012c] Zhang Y-J. 2012c. Image Engineering (2): Image Analysis (3rd Ed.). Tsinghua University Press, Beijing.

[Zhang 2012d] Zhang Y-J. 2012d. Image Engineering (3): Image Understanding (3rd Ed.). Tsinghua University Press, Beijing.

[Zhang 2013a] Zhang Y-J. 2013a. Image engineering in China: 2012. Journal of Image and Graphics, 18(5): 483–492.

[Zhang 2013b] Zhang Y-J. 2013b. Image Engineering (3rd Ed.). Tsinghua University Press, Beijing.

[Zhang 2014] Zhang Y-J. 2014. Image engineering in China: 2013. Journal of Image and Graphics, 19(5): 649–658.

[Zhang 2015a] Zhang Y-J. 2015a. Image engineering in China: 2014. Journal of Image and Graphics, 20(5): 585–598.

[Zhang 2015b] Zhang Y-J. 2015b. Image inpainting as an evolving topic in image engineering. In: Encyclopedia of Information Science and Technology (3rd Ed.), Information Science Reference, Hershey PA, 1283–1293 (Chapter 122).

[Zhang 2015c] Zhang Y-J. 2015c. Statistics on image engineering literatures. In: Encyclopedia of Information Science and Technology (3rd Ed.), Information Science Reference, Hershey PA, 6030–6040 (Chapter 595).

[Zhang 2016] Zhang Y-J. 2016. Image engineering in China: 2015. Journal of Image and Graphics, 21(5): 533–543.

[Zhang 2017] Zhang Y-J. 2017. Image engineering in China: 2016. Journal of Image and Graphics, 22(5): 563–573.

[Zhang et al. 2003] Zhang G C, Wang R D, Zhang Y-J. 2003. Digital image information hiding technology based on iterative blending. Chinese Journal of Computers, 26(5): 569–574.

[Zhang et al. 1999] Zhang N, Zhang Y-J, Liu Q D, et al. 1999. Method for estimating lossless image compression bound. IEE EL, 35(22): 1931–1932.

[Zhang et al. 2000] Zhang N, Lin X G, Zhang Y-J. 2000. L-infinity constrained micro noise filtering for high fidelity image compression. SPIE, 4067(1): 576–585.

[Zhang et al. 2015] Zheng Y, Shen B, Yang, XF, et al. 2015. A locality preserving approach for kernel PCA. Proc. 8ICIG (LNCS 9217), Zhang Y J (ed.), Springer, Berlin, 121–135.

Index

www.ingramcontent.com/pod-product-compliance
Lightning Source LLC
Chambersburg PA
CBHW061344210326
41598CB00035B/5877